LONDON MATHEMATICAL SOCIETY LECTURE NOT

Managing Editor: Professor J.W.S. Cassels, Department of Pure Math
University of Cambridge, 16 Mill Lane, Cambridge CB2 1SB, England

The titles below are available from booksellers, or, in case of difficulty, from Cambridge University Press.

34 Representation theory of Lie groups, M.F. ATIYAH *et al*
46 p-adic analysis: a short course on recent work, N. KOBLITZ
50 Commutator calculus and groups of homotopy classes, H.J. BAUES
59 Applicable differential geometry, M. CRAMPIN & F.A.E. PIRANI
66 Several complex variables and complex manifolds II, M.J. FIELD
69 Representation theory, I.M. GELFAND *et al*
76 Spectral theory of linear differential operators and comparison algebras, H.O. CORDES
77 Isolated singular points on complete intersections, E.J.N. LOOIJENGA
83 Homogeneous structures on Riemannian manifolds, F. TRICERRI & L. VANHECKE
86 Topological topics, I.M. JAMES (ed)
87 Surveys in set theory, A.R.D. MATHIAS (ed)
88 FPF ring theory, C. FAITH & S. PAGE
89 An F-space sampler, N.J. KALTON, N.T. PECK & J.W. ROBERTS
90 Polytopes and symmetry, S.A. ROBERTSON
92 Representation of rings over skew fields, A.H. SCHOFIELD
93 Aspects of topology, I.M. JAMES & E.H. KRONHEIMER (eds)
94 Representations of general linear groups, G.D. JAMES
95 Low-dimensional topology 1982, R.A. FENN (ed)
96 Diophantine equations over function fields, R.C. MASON
97 Varieties of constructive mathematics, D.S. BRIDGES & F. RICHMAN
98 Localization in Noetherian rings, A.V. JATEGAONKAR
99 Methods of differential geometry in algebraic topology, M. KAROUBI & C. LERUSTE
100 Stopping time techniques for analysts and probabilists, L. EGGHE
104 Elliptic structures on 3-manifolds, C.B. THOMAS
105 A local spectral theory for closed operators, I. ERDELYI & WANG SHENGWANG
107 Compactification of Siegel moduli schemes, C-L. CHAI
108 Some topics in graph theory, H.P. YAP
109 Diophantine analysis, J. LOXTON & A. VAN DER POORTEN (eds)
110 An introduction to surreal numbers, H. GONSHOR
113 Lectures on the asymptotic theory of ideals, D. REES
114 Lectures on Bochner-Riesz means, K.M. DAVIS & Y-C. CHANG
115 An introduction to independence for analysts, H.G. DALES & W.H. WOODIN
116 Representations of algebras, P.J. WEBB (ed)
118 Skew linear groups, M. SHIRVANI & B. WEHRFRITZ
119 Triangulated categories in the representation theory of finite-dimensional algebras, D. HAPPEL
121 Proceedings of *Groups - St Andrews 1985*, E. ROBERTSON & C. CAMPBELL (eds)
122 Non-classical continuum mechanics, R.J. KNOPS & A.A. LACEY (eds)
125 Commutator theory for congruence modular varieties, R. FREESE & R. MCKENZIE
126 Van der Corput's method of exponential sums, S.W. GRAHAM & G. KOLESNIK
128 Descriptive set theory and the structure of sets of uniqueness, A.S. KECHRIS & A. LOUVEAU
129 The subgroup structure of the finite classical groups, P.B. KLEIDMAN & M.W. LIEBECK
130 Model theory and modules, M. PREST
131 Algebraic, extremal & metric combinatorics, M-M. DEZA, P. FRANKL & I.G. ROSENBERG (eds)
132 Whitehead groups of finite groups, ROBERT OLIVER
133 Linear algebraic monoids, MOHAN S. PUTCHA
134 Number theory and dynamical systems, M. DODSON & J. VICKERS (eds)
135 Operator algebras and applications, 1, D. EVANS & M. TAKESAKI (eds)
136 Operator algebras and applications, 2, D. EVANS & M. TAKESAKI (eds)
137 Analysis at Urbana, I, E. BERKSON, T. PECK, & J. UHL (eds)
138 Analysis at Urbana, II, E. BERKSON, T. PECK, & J. UHL (eds)
139 Advances in homotopy theory, S. SALAMON, B. STEER & W. SUTHERLAND (eds)
140 Geometric aspects of Banach spaces, E.M. PEINADOR & A. RODES (eds)
141 Surveys in combinatorics 1989, J. SIEMONS (ed)
144 Introduction to uniform spaces, I.M. JAMES
145 Homological questions in local algebra, JAN R. STROOKER
146 Cohen-Macaulay modules over Cohen-Macaulay rings, Y. YOSHINO
147 Continuous and discrete modules, S.H. MOHAMED & B.J. MÜLLER
148 Helices and vector bundles, A.N. RUDAKOV *et al*
149 Solitons, nonlinear evolution equations and inverse scattering, M. ABLOWITZ & P. CLARKSON
150 Geometry of low-dimensional manifolds 1, S. DONALDSON & C.B. THOMAS (eds)

151 Geometry of low-dimensional manifolds 2, S. DONALDSON & C.B. THOMAS (eds)
152 Oligomorphic permutation groups, P. CAMERON
153 L-functions and arithmetic, J. COATES & M.J. TAYLOR (eds)
154 Number theory and cryptography, J. LOXTON (ed)
155 Classification theories of polarized varieties, TAKAO FUJITA
156 Twistors in mathematics and physics, T.N. BAILEY & R.J. BASTON (eds)
157 Analytic pro-*p* groups, J.D. DIXON, M.P.F. DU SAUTOY, A. MANN & D. SEGAL
158 Geometry of Banach spaces, P.F.X. MÜLLER & W. SCHACHERMAYER (eds)
159 Groups St Andrews 1989 volume 1, C.M. CAMPBELL & E.F. ROBERTSON (eds)
160 Groups St Andrews 1989 volume 2, C.M. CAMPBELL & E.F. ROBERTSON (eds)
161 Lectures on block theory, BURKHARD KÜLSHAMMER
162 Harmonic analysis and representation theory, A. FIGA-TALAMANCA & C. NEBBIA
163 Topics in varieties of group representations, S.M. VOVSI
164 Quasi-symmetric designs, M.S. SHRIKANDE & S.S. SANE
165 Groups, combinatorics & geometry, M.W. LIEBECK & J. SAXL (eds)
166 Surveys in combinatorics, 1991, A.D. KEEDWELL (ed)
167 Stochastic analysis, M.T. BARLOW & N.H. BINGHAM (eds)
168 Representations of algebras, H. TACHIKAWA & S. BRENNER (eds)
169 Boolean function complexity, M.S. PATERSON (ed)
170 Manifolds with singularities and the Adams-Novikov spectral sequence, B. BOTVINNIK
171 Squares, A.R. RAJWADE
172 Algebraic varieties, GEORGE R. KEMPF
173 Discrete groups and geometry, W.J. HARVEY & C. MACLACHLAN (eds)
174 Lectures on mechanics, J.E. MARSDEN
175 Adams memorial symposium on algebraic topology 1, N. RAY & G. WALKER (eds)
176 Adams memorial symposium on algebraic topology 2, N. RAY & G. WALKER (eds)
177 Applications of categories in computer science, M. FOURMAN, P. JOHNSTONE, & A. PITTS (eds)
178 Lower K- and L-theory, A. RANICKI
179 Complex projective geometry, G. ELLINGSRUD *et al*
180 Lectures on ergodic theory and Pesin theory on compact manifolds, M. POLLICOTT
181 Geometric group theory I, G.A. NIBLO & M.A. ROLLER (eds)
182 Geometric group theory II, G.A. NIBLO & M.A. ROLLER (eds)
183 Shintani zeta functions, A. YUKIE
184 Arithmetical functions, W. SCHWARZ & J. SPILKER
185 Representations of solvable groups, O. MANZ & T.R. WOLF
186 Complexity: knots, colourings and counting, D.J.A. WELSH
187 Surveys in combinatorics, 1993, K. WALKER (ed)
188 Local analysis for the odd order theorem, H. BENDER & G. GLAUBERMAN
189 Locally presentable and accessible categories, J. ADAMEK & J. ROSICKY
190 Polynomial invariants of finite groups, D.J. BENSON
191 Finite geometry and combinatorics, F. DE CLERCK *et al*
192 Symplectic geometry, D. SALAMON (ed)
193 Computer algebra and differential equations, E. TOURNIER (ed)
194 Independent random variables and rearrangement invariant spaces, M. BRAVERMAN
195 Arithmetic of blowup algebras, WOLMER VASCONCELOS
196 Microlocal analysis for differential operators, A. GRIGIS & J. SJÖSTRAND
197 Two-dimensional homotopy and combinatorial group theory, C. HOG-ANGELONI, W. METZLER & A.J. SIERADSKI (eds)
198 The algebraic characterization of geometric 4-manifolds, J.A. HILLMAN
199 Invariant potential theory in the unit ball of C^n, MANFRED STOLL
200 The Grothendieck theory of dessins d'enfant, L. SCHNEPS (ed)
201 Singularities, JEAN-PAUL BRASSELET (ed)
202 The technique of pseudodifferential operators, H.O. CORDES
203 Hochschild cohomology of von Neumann algebras, A. SINCLAIR & R. SMITH
204 Combinatorial and geometric group theory, A.J. DUNCAN, N.D. GILBERT & J. HOWIE (eds)
205 Ergodic theory and its connections with harmonic analysis, K. PETERSEN & I. SALAMA (eds)
206 Lectures on noncommutative geometry, J. MADORE
207 Groups of Lie type and their geometries, W.M. KANTOR & L. DI MARTINO (eds)
208 Vector bundles in algebraic geometry, N.J. HITCHIN, P. NEWSTEAD & W.M. OXBURY (eds)
209 Arithmetic of diagonal hypersurfaces over finite fields, F.Q. GOUVÊA & N. YUI
210 Hilbert C*-modules, E.C. LANCE
211 Groups 93 Galway / St Andrews I, C.R. CAMPBELL *et al*
212 Groups 93 Galway / St Andrews II, C.R. CAMPBELL *et al*
215 Number theory, S. DAVID (ed)
216 Stochastic partial differential equations, A. ETHERIDGE (ed)
221 Harmonic approximation, S.J. GARDINER

London Mathematical Society Lecture Note Series. 221

Harmonic Approximation

Stephen J. Gardiner
University College, Dublin

Published by the Press Syndicate of the University of Cambridge
The Pitt Building, Trumpington Street, Cambridge CB2 1RP
40 West 20th Street, New York, NY 10011-4211, USA
10 Stamford Road, Oakleigh, Melbourne 3166, Australia

First published 1995

Printed in Great Britain at the University Press, Cambridge

Library of Congress cataloging in publication data available

British Library cataloguing in publication data available

ISBN 0 521 49799 X paperback

To Lindsey

Table of Contents

Preface

The year 1885 has a special significance in the history of approximation theory. It was then that Weierstrass published his famous result which says that a continuous function on a closed bounded interval of the real line can be uniformly approximated by polynomials. The same year saw the birth of holomorphic approximation in the celebrated paper of Runge [Run]. Given an open set Ω in the complex plane $\mathbf{C}$, which compact subsets K have the property that any holomorphic function defined on a neighbourhood of K can be uniformly approximated on K by functions holomorphic on Ω? Runge's Theorem supplies the answer: precisely the sets K such that $\Omega \backslash K$ has no components which are relatively compact in Ω. Since Runge's original work holomorphic approximation has developed into a significant research area. We mention particularly the contributions of Carleman [CarT], Alice Roth [Rot1], [Rot3], Mergelyan [Mer], Arakelyan [Ara1] and Nersesyan [Ner]. A helpful account of these and other results can be found in the book by Gaier [Gai]. The purpose of these notes is to give a corresponding account of the theory of harmonic approximation in Euclidean space $\mathbf{R}^n$ $(n \geq 2)$.

The starting point in the history of harmonic approximation is not as easy to identify. In the case of approximation in higher dimensions, the paper of Walsh [Wal] in 1929 seems a reasonable choice, but for approximation in the plane mention must also be made of work of Lebesgue [Leb] in 1907. Which compact sets K in $\mathbf{R}^n$ have the property that any harmonic function defined on a neighbourhood of K can be uniformly approximated on K by harmonic polynomials? Walsh's Theorem tells us that, if $\mathbf{R}^n \backslash K$ is connected, then such approximation is always possible. However, unlike the case of holomorphic approximation, the converse to this statement is false. The characterization of compact sets K with the above approximation property is rather more delicate and involves the potential theoretic notion of "thin sets": denoting by $\widehat{K}$ the union of K with the bounded components of $\mathbf{R}^n \backslash K$, the relevant condition is that $\mathbf{R}^n \backslash \widehat{K}$ and $\mathbf{R}^n \backslash K$ must be thin at the same points of K (see [Gar3]).

The literature on harmonic approximation has also become extensive. There was a period of significant development in the 1940's, due particularly to Keldyš [Kel], Landkof (see [Lan] and the references given there), Brelot [Brel] and Deny [Den1], [Den2]. Most of this work was concerned with the question of approximation by locally- (rather than by globally-) defined harmonic functions. Which compact sets K in $\mathbf{R}^n$ have the property that any function which is continuous on K and harmonic on the interior K° can be uniformly approximated by functions harmonic on a neighbourhood of K? The answer here also involves thin sets: $\mathbf{R}^n \backslash K$ and $\mathbf{R}^n \backslash K^\circ$ must be thin at the same points.

Until comparatively recently most of the work was in terms of approximation on *compact* sets. This changed in the early 1980's due to two papers by Gauthier, Goldstein and Ow [GGO1], [GGO2]. Inspired by work of Alice Roth [Rot3] in the holomorphic case, they developed a technique of "fusing" two harmonic functions which are close in value on a certain set. As a result they obtained, in particular, a generalization of Walsh's Theorem to the case of approximation on closed (but not necessarily bounded) sets E in $\mathbf{R}^n$: if $(\mathbf{R}^n)^* \backslash E$ is connected and locally connected (where $(\mathbf{R}^n)^*$ denotes the one-point compactification $\mathbf{R}^n \cup \{\infty\}$ of $\mathbf{R}^n$), then functions harmonic on an open set containing E can be uniformly approximated on E by functions harmonic on all of $\mathbf{R}^n$. As was the case with Walsh's Theorem, the above hypotheses concerning connectedness are sufficient, but not necessary for this type of approximation to be possible. A complete characterization of sets E which possess this approximation property has recently been given by the author [Gar3]. Indeed, a recent period of rapid development, due to several authors, has brought a new coherence and substance to the whole subject of harmonic approximation. The purpose of these lecture notes is to give an organised account of harmonic approximation which includes many of these new results.

The plan of these notes is as follows. We assume that the reader is familiar with an introductory text on potential theory such as the book by Helms [Helm], but for convenience we collect together in a preliminary chapter some particularly relevant facts concerning thin sets. Uniform harmonic approximation on compact sets, and then on relatively closed sets, is dealt with in Chapters 1 and 3 respectively. The contents of Chapter 3 and much of the subsequent work rely on a fusion result derived in Chapter 2. We then turn our attention to the question of better-than-uniform approximation: that is, can it be arranged that the error in our approximation decays to 0 as we approach "infinity"? In the holomorphic case one such famous result is due to Carleman [CarT]. He showed that, given any continuous functions $f : \mathbf{R} \to \mathbf{C}$ and $\epsilon : \mathbf{R} \to (0, 1]$, there exists an entire function g such that $|g - f| < \epsilon$ on $\mathbf{R}$. Various results of this type are presented in

Chapters 3-5. Chapter 6 contains analogous and related results concerning the approximation and extension of superharmonic functions.

Finally, one of the most rewarding aspects of the whole subject is its potential for applications, sometimes surprising ones. A selection of these applications will be presented in Chapters 7 and 8. We mention two of them here briefly. The first, which appears in Chapter 7, concerns the Dirichlet problem on an unbounded open set Ω in $\mathbf{R}^n$. Given any continuous function f on the (non-compact) boundary $\partial\Omega$, can one find a harmonic function h_f on Ω which satisfies $h_f(X) \to f(Y)$ as $X \to Y$ for all regular boundary points Y? R. Nevanlinna [Nev] showed in 1925 that this was possible in the case where Ω is a half-plane, and several authors subsequently considered the question further. Results in Chapter 3 will be used to give a complete topological characterization of the sets Ω in which such a Dirichlet problem can always be solved. The second application that we mention here concerns the Radon transform, which itself has applications to tomography. Let f be a real- or complex-valued function on $\mathbf{R}^n$ such that f is integrable on each $(n-1)$-dimensional hyperplane P in $\mathbf{R}^n$. The Radon transform $\widehat{f}$ is defined on the collection $\mathcal{P}^{(n)}$ of all such hyperplanes by $\widehat{f}(P) = \int_P f$ for each P in $\mathcal{P}^{(n)}$, where the integration is with respect to $(n-1)$-dimensional Lebesgue measure on P. An old question concerning the Radon transform was whether there exists a non-constant continuous function f such that $\widehat{f} \equiv 0$ on $\mathcal{P}^{(n)}$. In Chapter 8 we show how Armitage and Goldstein [AG2] used an approximation theorem from Chapter 5 to answer this question: surprisingly, there even exists a non-constant *harmonic* function on $\mathbf{R}^n$ with this property! (When $n = 2$ the question had previously been settled by Zalcman [Zal] using holomorphic approximation.)

These notes began to take shape while I was on sabbatical at McGill University during the calendar year 1992. Some of the early material formed part of an advanced course of lectures entitled *Thin sets and their applications* which I gave there during the Fall Semester. I would like to thank the Department of Mathematics and Statistics at McGill, and especially Prof. K. N. GowriSankaran, for making my visit possible. I am grateful also to Professors David Armitage, Paul Gauthier, Maciej Klimek and Ivan Netuka for their comments on a first draft of these notes. Finally I wish to thank Siobhán Purcell for her assistance in preparing the camera-ready copy and Professor Mícheál Ó Searcóid for the preparation of the diagrams.

0 Review of Thin Sets

The purpose of this preliminary chapter is to give a brief review of facts concerning thin sets which are particularly relevant to harmonic approximation. No proofs will be given, but appropriate references to the books by Helms [Helm] and Doob [Doo] will be supplied.

0.1 Introduction

We use $\overline{A}$, ∂A and A° to denote respectively the closure, boundary and interior of a set A in Euclidean space $\mathbf{R}^n$ $(n \geq 2)$, denote by $|X|$ the Euclidean norm of a point X, and denote by $B(X,r)$ the open ball of centre X and radius r. Also, we define $\phi_n : [0, +\infty) \to \mathbf{R} \cup \{+\infty\}$ by $\phi_2(t) = \log(1/t)$, or $\phi_n(t) = t^{2-n}$ if $n \geq 3$. (We interpret $\phi_n(0)$ as $+\infty$ in either case.) Let Ω be an open set in $\mathbf{R}^n$. A function u on Ω, taking values in $(-\infty, +\infty]$, is called *superharmonic* if:

(i) $u \not\equiv +\infty$ on any component of Ω;
(ii) u is lower semicontinuous, i.e. the set $\{X \in \Omega : u(X) > a\}$ is open for each real number a; and
(iii) $u(X) \geq \mathcal{M}(u; X, r)$ whenever $\overline{B(X,r)} \subseteq \Omega$, where $\mathcal{M}(u; X, r)$ denotes the mean value of u over the sphere $\partial B(X, r)$.

A fundamental example of a superharmonic function on $\mathbf{R}^n$ is given by $\phi_n(|X|)$, which is harmonic on $\mathbf{R}^n \backslash \{O\}$, and which takes the value $+\infty$ at the origin O. It is a consequence of the above definition that a superharmonic function can take the value $+\infty$ only on a rather small set of points. A set A is called *polar* if there is a superharmonic function u on $\mathbf{R}^n$ such that $A \subseteq \{X : u(X) = +\infty\}$. Thus, for example, the set $\{0\}^m \times \mathbf{R}^{n-m}$ is polar if $m \in \{2, 3, \ldots, n\}$, as can be seen by considering the superharmonic function

$$X \mapsto \phi_m\big((x_1^2 + x_2^2 + \ldots + x_m^2)^{1/2}\big).$$

However, the hyperplane $\{0\} \times \mathbf{R}^{n-1}$ is not polar. Polar sets always have n-dimensional Lebesgue measure 0.

As the definition suggests, superharmonic functions need not be continuous, even in the extended sense of functions taking values in $[-\infty, +\infty]$. In fact, if $\{Y_k : k \in \mathbf{N}\}$ is a countable dense subset of some ball, where $\mathbf{N} = \{1, 2, \ldots\}$, then a highly discontinuous superharmonic function is defined by

$$u(X) = \sum_{k=1}^{\infty} 2^{-k} \phi_n(|X - Y_k|) \quad (X \in \mathbf{R}^n).$$

Nevertheless, it is possible to assert that, if u is a superharmonic function on Ω and $Y \in \Omega$, then

$$u(X) \to u(Y) \quad (X \to Y; X \notin E),$$

where the exceptional set E is, in a sense which becomes clear from Wiener's criterion in §0.5, "thin" at Y.

0.2 The Fine Topology

If T_1, T_2 are topologies on the same set and $T_2 \subseteq T_1$, then T_1 is said to be *finer* than T_2, and T_2 is said to be *coarser* than T_1. The *fine topology* of classical potential theory is the coarsest topology on $\mathbf{R}^n$ which makes every superharmonic function on $\mathbf{R}^n$ continuous. It is obtained by taking the intersection of all topologies which make the superharmonic functions continuous. The fine topology clearly contains all open balls, and hence the Euclidean topology. It is strictly finer than the Euclidean topology since, as we have seen, there exist superharmonic functions which are discontinuous with respect to the latter.

A set E is said to be *thin* at a point Y if Y is not a fine limit point of E, i.e. if there is a fine (topology) neighbourhood N of Y such that $E \backslash \{Y\}$ does not intersect N. The classical example of a thin set is the *Lebesgue spine* in $\mathbf{R}^3$ defined by

$$E_c = \{(x, y, z) : x > 0 \text{ and } y^2 + z^2 < \exp(-c/x)\} \quad (c > 0), \tag{0.1}$$

which is thin at O. A polar set is thin at every point of $\mathbf{R}^n$.

Theorem 0.A. *([Helm, 10.3], [Doo, 1.XI.2]) Let Y be a limit point of a set E. Then E is thin at Y if and only if there is a superharmonic function u on an open neighbourhood of Y such that*

$$\liminf_{X \to Y, X \in E} u(X) > u(Y).$$

(Throughout these notes, limit concepts for a function u do not involve the value of u at the point concerned.)

An open set Ω in $\mathbf{R}^n$ will be called *Greenian* if it possesses a Green function $G_\Omega(\cdot,\cdot)$. When $n \geq 3$ all open sets are Greenian. When $n = 2$ an open set Ω is Greenian if and only if its complement is a not a polar set (see [Helm, 8.33]). If u is a non-negative superharmonic function on a Greenian open set Ω such that the greatest harmonic minorant of u on Ω is the zero function, then u is called a *potential* on Ω. In this case there exists a unique (Borel) measure ν_u on Ω such that $u = G_\Omega \nu_u$ where

$$G_\Omega \nu_u(X) = \int_\Omega G_\Omega(Y,X)\,d\nu_u(Y) \quad (X \in \Omega).$$

Any non-negative superharmonic function u on Ω can be written as the sum of its greatest harmonic minorant and a potential of the above form. The measure ν_u is called the *Riesz measure* associated with u. It is given by $\nu_u = -c_n \Delta u$ in the sense of distributions, where $c_n^{-1} = \sigma_n \max\{n-2, 1\}$ and σ_n denotes the surface area of the unit sphere in $\mathbf{R}^n$.

Theorem 0.B. *([Doo, 1.XI.2], cf. [Helm, 10.4]) Let Ω be a Greenian open set and suppose that a set E is thin at a limit point Y in Ω. Then there is a potential u on Ω such that*

$$u(Y) < \liminf_{X \to Y, X \in E} u(X) = +\infty.$$

Corollary 0.C. *([Helm, 10.5]) If a Borel set E is thin at a point Y, then*

$$\frac{\sigma\big(\partial B(Y,r) \cap E\big)}{\sigma\big(\partial B(Y,r)\big)} \to 0 \qquad (r \to 0+),$$

where σ denotes surface area measure on $\partial B(Y,r)$.

Theorem 0.D. *([Helm, 10.14]) If a set E in $\mathbf{R}^2$ is thin at a point Y, then there are arbitrarily small positive values of r such that $\partial B(Y,r) \cap E = \emptyset$.*

Theorem 0.E. *([Helm, 10.9], [Doo, 1.XI.6]) Let $E \subseteq \mathbf{R}^n$. The set of points of E where E is thin forms a polar set.*

We note from Theorem 0.E that, since an $(n-1)$-dimensional hyperplane (or line, if $n = 2$) is non-polar, it cannot be thin at any of its constituent points, in view of its translational symmetries.

0.3 Reduced Functions and Thinness

Let Ω be a Greenian open set, let u be a non-negative superharmonic function on Ω, and let $E \subseteq \Omega$. The *réduite*, or *reduced function*, of u relative to E in Ω is defined by

$$R_u^E(X) = \inf\{v(X) : v \text{ is a superharmonic function on } \Omega, \\ v \geq 0 \text{ on } \Omega, v \geq u \text{ on } E\}$$

when $X \in \Omega$. Its lower regularization, that is

$$\widehat{R}_u^E(X) = \min\left\{\liminf_{Y \to X} R_u^E(Y), R_u^E(X)\right\} \qquad (X \in \Omega),$$

is called the *balayage*, or *regularized reduced function*, of u relative to E in Ω. The balayage is a superharmonic function on Ω. It is obvious that $u \geq R_u^E \geq \widehat{R}_u^E \geq 0$ on Ω, that $u = R_u^E$ on E, and that $u = R_u^E = \widehat{R}_u^E$ on E°. Some further properties of reduced functions are listed below:

(i) R_u^E (and hence also $\widehat{R}_u^E$) is harmonic on $\Omega \backslash \overline{E}$ ([Helm, 7.11]);

(ii) $\widehat{R}_u^E = R_u^E$ on $\Omega \backslash E$ ([Helm, 8.36]);

(iii) $\widehat{R}_u^E = R_u^E$ on Ω if E is open (cf. (ii));

(iv) $\widehat{R}_u^E$ differs from R_u^E at most on a polar set ([Helm, 7.39]);

(v) if F is a polar set, then $\widehat{R}_u^{E \backslash F} = \widehat{R}_u^E$ on Ω ([Helm, 8.37]);

(vi) if (E_k) is an increasing sequence of sets and $E = \cup_k E_k$, then $R_u^{E_k} \uparrow R_u^E$ and $\widehat{R}_u^{E_k} \uparrow \widehat{R}_u^E$ (cf. [Helm, 8.38]).

(All the above properties can also be found in [Doo, 1.VI.3].)

Theorem 0.F. *([Doo, 1.XI.10]) Let Ω be a Greenian open set. Then there is a bounded continuous potential $u^\#$ on Ω with the property that a set E is thin at a point Y in Ω if and only if $\widehat{R}_{u^\#}^E(Y) < u^\#(Y)$.*

0.4 Thin Sets and the Dirichlet Problem

We refer to [Helm, Chapters 8,9] and [Doo, 1.VIII] for accounts of the Perron-Wiener-Brelot solution to the Dirichlet problem on a Greenian open set Ω with boundary function $f : \partial^*\Omega \to [-\infty, +\infty]$. Here $\partial^*\Omega$ denotes $\partial\Omega$ if Ω is bounded, or $\partial\Omega \cup \{\infty\}$ if Ω is unbounded, and ∞ denotes the Alexandroff point for $\mathbf{R}^n$. We recall that a function u on Ω is said to be in the *upper* (resp. *lower*) *PWB class* if, on each component of Ω, either $u \equiv +\infty$ (resp. $u \equiv -\infty$) or u is superharmonic (resp. subharmonic) and bounded below (resp. above), and if

$$\liminf_{X \to Y} u(X) \geq f(Y) \qquad (\text{resp.} \limsup_{X \to Y} u(X) \leq f(Y)) \qquad (Y \in \partial^*\Omega).$$

Further, the infimum (resp. supremum) of the upper (resp. lower) PWB class is denoted by $\overline{H}_f^\Omega$ (resp. $\underline{H}_f^\Omega$). If $\overline{H}_f^\Omega$ and $\underline{H}_f^\Omega$ are identical and harmonic on Ω, then we denote them by H_f^Ω. In this case f is said to be *resolutive* for Ω, and H_f^Ω is called the *PWB solution* for f. A point Y in $\partial^*\Omega$ is called *regular* if $H_f^\Omega(X) \to f(X)$ as $X \to Y$ for every continuous function $f : \partial^*\Omega \to \mathbf{R}$. Otherwise Y is called *irregular*. The set Ω is called *regular* if every point in $\partial^*\Omega$ is regular. For each X in Ω there is a unique (Borel) measure $\mu_{\Omega,X}$ on $\partial^*\Omega$ such that

$$H_f^\Omega(X) = \int_{\partial^*\Omega} f(Y)\, d\mu_{\Omega,X}(Y) \qquad (X \in \Omega)$$

for every resolutive boundary function f. The measure $\mu_{\Omega,X}$ is called *harmonic measure* for Ω and X. If Ω is connected, then the class of Borel subsets of $\partial^*\Omega$ which have zero $\mu_{\Omega,X}$-measure is independent of X. In connection with the following results we emphasize that $\partial\Omega$ denotes the *Euclidean* boundary of Ω, and so does not include ∞ even if Ω is unbounded.

Theorem 0.G. *([Helm, 10.12], [Doo, 1.XI.12]) Let Ω be a Greenian open set and let $Y \in \partial\Omega$. Then Y is a regular boundary point for the Dirichlet problem on Ω if and only if $\mathbf{R}^n \backslash \Omega$ is not thin at Y.*

Theorem 0.H. *([Doo, 1.XI.13]) Let Ω be a connected Greenian open set. Then the set of points of $\partial\Omega$ at which Ω is thin forms a set of zero harmonic measure for Ω.*

We also record here the relationship between reduced functions and Dirichlet solutions.

Theorem 0.I. *([Helm, 9.25], [Doo,1.VIII.10]) Let Ω be a Greenian open set, let ω be an open subset of Ω, and let u be a positive superharmonic function on Ω. Then $R_u^{\Omega \backslash \omega} = H_{u'}^\omega$ on ω, where*

$$u'(X) = \begin{cases} u(X) & (X \in \Omega \cap \partial\omega) \\ 0 & (X \in \partial\Omega \cap \partial\omega; X = \infty \textit{ if } \omega \textit{ is unbounded}). \end{cases}$$

0.5 Wiener's Criterion

Let Ω be an open set in $\mathbf{R}^n$ with Green function $G_\Omega(\,\cdot\,,\,\cdot\,)$, let $E \subseteq \mathbf{R}^n$ and $Y \in \Omega$, and let α denote a fixed number satisfying $\alpha > 1$. For each positive integer k we define

$$E_k = \{X \in E : \alpha^k \leq \phi_n(|X - Y|) \leq \alpha^{k+1}\}.$$

Also, we choose k' such that the closed ball $\{X : \alpha^{k'} \leq \phi_n(|X - Y|)\}$ is contained in Ω. In what follows, $\mathcal{C}^*$ denotes outer capacity with respect to Ω (see [Helm, Chap. 7] or [Doo, 1.XIII]). When $n \geq 3$ we may take $\Omega = \mathbf{R}^n$, in which case $\mathcal{C}^*$ is outer Newtonian capacity.

Theorem 0.J. *([Helm, 10.21], [Doo, 1.XI.3 and 1.XIII.17]) The following are equivalent:*

(i) E is thin at Y;

(ii) $\sum_{k=k'}^{\infty} \alpha^k \mathcal{C}^*(E_k) < +\infty$ (Wiener's criterion);

(iii) $\displaystyle\int_{\alpha^{k'}}^{\infty} \mathcal{C}^*(\{X \in E : \phi_n(|X - Y|) > t\})\, dt < +\infty$;

(iv) $\widehat{R}^E_{G_\Omega(Y,\,\cdot\,)} \neq G_\Omega(Y,\,\cdot\,)$ (unequal as functions).

Theorem 0.K. *([Doo, 1.XI.4]) Let Ω be a Greenian open set, and let u be a positive superharmonic function on Ω with associated Riesz measure ν_u. Then $u/G_\Omega(Y,\,\cdot\,)$ has fine limit $\nu_u(\{Y\})$ at Y.*

1 Approximation on Compact Sets

1.1 Introduction

If Ω is an open set in $\mathbf{C}$ or $\mathbf{R}^n (n \geq 2)$, then we will use Ω^* to denote the Alexandroff, or one-point, compactification of Ω, and will use $\mathcal{A}$ to denote the ideal point. Thus $\Omega^* = \Omega \cup \{\mathcal{A}\}$, and a set A is open in Ω^* if either A is an open subset of Ω or $A = \Omega^* \backslash K$, where K is a compact subset of Ω. In the special case where Ω is $\mathbf{C}$ or $\mathbf{R}^n$ we continue to write ∞ for $\mathcal{A}$.

If A is a subset of $\mathbf{C}$, we denote by $\mathrm{Hol}(A)$ the collection of all functions which are holomorphic on an open set containing A. Historically the following result (essentially in [Run]; cf. [Con, pp.198, 201]) can be regarded as the starting point of the theory of holomorphic approximation.

Runge's Theorem (1885). *Let Ω be an open subset of $\mathbf{C}$ and K be a compact subset of Ω. The following are equivalent:*
(a) for each f in $\mathrm{Hol}(K)$ and each positive number ϵ, there exists g in $\mathrm{Hol}(\Omega)$ such that $|g - f| < \epsilon$ on K;
(b) $\Omega^* \backslash K$ is connected.

Condition (b) above is equivalent to asserting that no component of $\Omega \backslash K$ is relatively compact in Ω. Also, when $\Omega = \mathbf{C}$, this condition is clearly equivalent to saying that $\mathbf{C} \backslash K$ is connected.

We record below one further important development in the theory of holomorphic approximation, which deals with approximation of a much larger class of functions on a given compact set K. It can be found in [Mer] or [Rud, Chap. 20]. As usual we denote by $C(A)$ the collection of all complex- (or real-, depending on the context) valued continuous functions on a set A.

Mergelyan's Theorem (1952). *Let K be a compact set in $\mathbf{C}$. The following are equivalent:*
(a) for each f in $C(K) \cap \mathrm{Hol}(K^o)$ and each positive number ϵ, there is an

entire function (and hence, by suitably truncating the Taylor series, a polynomial) g such that $|g - f| < \epsilon$ on K;
(b) $\mathbf{C}\backslash K$ is connected.

Turning now to the history of harmonic approximation in $\mathbf{R}^n (n \geq 2)$, we take as our starting point the following result [Wal, p.541].

Walsh's Theorem (1929). *Let K be a compact set in $\mathbf{R}^n$ such that $\mathbf{R}^n\backslash K$ is connected. Then, for each function u which is harmonic on an open set containing K and each positive number ϵ, there is a harmonic polynomial v such that $|v - u| < \epsilon$ on K.*

Important progress concerning uniform harmonic approximation was made in the 1940's, in which connection we mention particularly the contributions by Keldyš [Kel], Landkof (see the references in [Lan]), Brelot [Brel], and Deny [Den1], [Den2]. However, somewhat surprisingly, the analogue of Runge's Theorem (as stated above) for harmonic approximation in $\mathbf{R}^n$ was obtained rather more recently. In this chapter we present analogues of both Runge's Theorem and Mergelyan's Theorem for harmonic functions. First, however, we deal with the question of local harmonic approximation, i.e. approximation by harmonic functions defined merely on some neighbourhood of a given compact set.

1.2 Local Approximation on Compact Sets with Empty Interior

In this section and the next we will be concerned with uniform approximation of functions on a compact set K in $\mathbf{R}^n$ by functions harmonic on a neighbourhood of K. Clearly the functions to be approximated must be continuous on K and harmonic on the interior K°. It will be convenient to denote by $\mathcal{H}(A)$ the collection of all functions which are harmonic on some open set containing A. The question we are concerned with is this. Which compact sets K have the property that every u in $C(K) \cap \mathcal{H}(K^\circ)$ can be uniformly approximated by functions in $\mathcal{H}(K)$? This question simplifies if we restrict our attention to compact sets K with empty interior, so this special case will be treated first. In §1.3 the more general question will be dealt with. Of course, if $K^\circ = \emptyset$, then we are approximating arbitrary continuous functions on K.

Theorem 1.1. *Let K be a compact subset of $\mathbf{R}^n$ such that $K^\circ = \emptyset$. The following are equivalent:*

(a) for each f in $C(K)$ and each positive number ϵ, there exists h in $\mathcal{H}(K)$ such that $|h - f| < \epsilon$ on K;
(b) $\mathbf{R}^n \backslash K$ is nowhere thin.

Before proving this theorem we present an example of a compact set K which has empty interior yet fails to satisfy condition (b).

Example 1.2. Let $\{Y_k : k \in \mathbf{N}\}$ be a dense subset of $[0,1]^{n-1} \times (0,1]$, and define

$$u(X) = \sum_{k=1}^{\infty} 2^{-k} \phi_n(|X - Y_k|) \qquad (X \in \mathbf{R}^n),$$

$$E = \left([0,1]^{n-1} \times \{0\}\right) \cup \left\{(X', x_n) \in [0,1]^{n-1} \times (0,1] : u(X', x_n) \leq \phi_n(x_n)\right\}$$

and

$$K = \left\{(x_1, \ldots, x_n) \in \mathbf{R}^n : (x_1, \ldots, x_{n-1}, |x_n|) \in E\right\}.$$

Then u is a superharmonic function on $\mathbf{R}^n$, and the lower semicontinuity of u ensures that E is closed. Thus K is compact and, because $u(Y_k) = +\infty$ for each k, the interior K° is empty. (The set K is an example of what is sometimes called a "Swiss cheese": cf. [Rot1, p.96].) If $Z \in (0,1)^{n-1} \times \{0\}$ and $X = (X', x_n)$, where $x_n > 0$, then

$$u(X) > \phi_n(x_n) \geq \phi_n(|X - Z|) \qquad (X \in (0,1)^n \backslash E).$$

Since the Riesz measure associated with u does not charge $\{Z\}$, it follows easily from Theorem 0.K that $(0,1)^n \backslash E$ is thin at Z, and hence $\mathbf{R}^n \backslash K$ is thin at Z.

In both parts of the proof below, K will denote a compact set with empty interior, B will be a fixed open ball which contains K, and reductions will be with respect to superharmonic functions on B. Also, we define

$$U_m = \{X \in \mathbf{R}^n : \text{dist}(X, K) < 1/m\} \qquad (m \in \mathbf{N}).$$

Proof that (b) implies (a). Suppose that $\mathbf{R}^n \backslash K$ is nowhere thin, let $f \in C(K)$ and $\epsilon > 0$. There exist (see [Helm, 8.10]) positive continuous superharmonic functions u_1, u_2 on B such that

$$\left|f - (u_1 - u_2)\right| < \epsilon/2 \text{ on } K. \tag{1.1}$$

We know that

$$R_{u_k}^{B \backslash U_m}(X) \uparrow R_{u_k}^{B \backslash K}(X) \qquad (X \in B; k = 1, 2; m \to \infty). \tag{1.2}$$

Now let v_k be a positive superharmonic function on B such that $v_k \geq u_k$ on $B \backslash K$, where $k \in \{1, 2\}$. It follows by fine continuity that $v_k - u_k \geq 0$ on K, since $B \backslash K$ is non-thin at each X in K and so every fine neighbourhood of such a point X meets $B \backslash K$. Thus $v_k \geq u_k$ on B and hence

$$u_k = R_{u_k}^{B \backslash K} \qquad \text{on } B.$$

From (1.2) we see that

$$R_{u_k}^{B \backslash U_m}(X) \uparrow u_k(X) \qquad (X \in K; k = 1, 2; m \to \infty)$$

and, since u_k is continuous and K is compact, Dini's theorem shows that this convergence is uniform on K. Thus there exists m' such that

$$u_k(X) \geq R_{u_k}^{B \backslash U_{m'}}(X) > u_k(X) - \epsilon/2 \qquad (X \in K; k = 1, 2). \tag{1.3}$$

It follows from (1.1) and (1.3) that

$$f - R_{u_1}^{B \backslash U_{m'}} + R_{u_2}^{B \backslash U_{m'}} < f - u_1 + \epsilon/2 + u_2 < \epsilon \qquad \text{on } K,$$

and similarly

$$f - R_{u_1}^{B \backslash U_{m'}} + R_{u_2}^{B \backslash U_{m'}} > -\epsilon \qquad \text{on } K.$$

Since the above reduced functions are harmonic on the open set $U_{m'}$ which contains K, the argument is complete.

Proof that (a) implies (b). Suppose that condition (a) of Theorem 1.1 holds, let $u^{\#}$ be the bounded continuous potential on B described in Theorem 0.F, and let $\epsilon > 0$. By hypothesis there exists h_ϵ in $\mathcal{H}(K)$ such that

$$|h_\epsilon - u^{\#}| < \epsilon \qquad \text{on } K. \tag{1.4}$$

By continuity the above inequality remains true on the open set U_m for all sufficiently large m. Thus, solving the Dirichlet problem in U_m, we obtain

$$-\epsilon \leq h_\epsilon - H_{u^{\#}}^{U_m} = h_\epsilon - R_{u^{\#}}^{B \backslash U_m} \leq \epsilon \qquad \text{on } K$$

(See Theorem 0.I) Letting $m \to \infty$, it follows that

$$\left| h_\epsilon - R_{u^{\#}}^{B \backslash K} \right| \leq \epsilon \qquad \text{on } K.$$

Combining this with (1.4) we see that

$$\left| u^{\#} - \widehat{R}_{u^{\#}}^{B \backslash K} \right| = \left| u^{\#} - R_{u^{\#}}^{B \backslash K} \right| < 2\epsilon \qquad \text{on } K.$$

Since ϵ can be arbitrarily small,

$$u^{\#} = \widehat{R}_{u^{\#}}^{B \backslash K} \qquad \text{on } K.$$

From Theorem 0.F we can conclude that $B \backslash K$, and hence $\mathbf{R}^n \backslash K$, is not thin at any point of K. Certainly $\mathbf{R}^n \backslash K$ is not thin at any point of $\mathbf{R}^n \backslash K$, so condition (b) of the theorem is established.

1.3 Local Harmonic Approximation

We return now to the question posed at the beginning of §1.2. Which compact sets K have the property that every u in $C(K) \cap \mathcal{H}(K^\circ)$ can be uniformly approximated by functions in $\mathcal{H}(K)$? Thus we are now dropping the additional assumption that $K^\circ = \emptyset$. This question is completely answered by the following result which clearly includes Theorem 1.1.

Theorem 1.3. *Let K be a compact subset of $\mathbf{R}^n$. The following are equivalent:*
(a) for each u in $C(K) \cap \mathcal{H}(K^\circ)$ and each positive number ϵ, there exists v in $\mathcal{H}(K)$ such that $|v - u| < \epsilon$ on K;
(b) $\mathbf{R}^n \backslash K$ and $\mathbf{R}^n \backslash K^\circ$ are thin at the same points (of K).

Example 1.4. (i) Condition (b) above fails to hold for the compact set K of Example 1.2. An example of a compact set with non-empty interior which fails to satisfy condition (b) can be obtained by defining $K = E \cup ([0,1]^{n-1} \times [-1,0])$, where E is as in Example 1.2.

(ii) In $\mathbf{R}^3$ let E_c be the Lebesgue spine defined in equation (0.1), and let $K = \overline{B(O,1)} \backslash E_2$. Then $\mathbf{R}^3 \backslash K$ is thin at O, and it is easy to see that O is the only point of ∂K at which $\mathbf{R}^3 \backslash K$ is thin. Further, $B(0,1) \backslash K^\circ \subset E_1 \cup \{0\}$, so $\mathbf{R}^3 \backslash K^\circ$ is also thin at O. Thus $\mathbf{R}^3 \backslash K$ and $\mathbf{R}^3 \backslash K^\circ$ are thin at precisely the same points (namely, points of $K^\circ \cup \{O\}$), and so condition (b) holds for this choice of K, even though K° has an irregular boundary point.

The proof that we shall give for the implication "(b) $\Rightarrow$ (a)" actually requires only the following slightly weaker hypothesis:

(c) $\{X \in K : \mathbf{R}^n \backslash K$ is thin at $X\} \cap \{X \in K : \mathbf{R}^n \backslash K^\circ$ is not thin at $X\}$ *is a polar set.*

Thus (c) is an additional equivalent condition in Theorem 1.3. If we take capacities relative to a Greenian open set Ω which contains K, then a further equivalent condition is:

(d) $\mathcal{C}(W \backslash K) = \mathcal{C}(W \backslash K^\circ)$ for every relatively compact open subset W of Ω.

To see this, we observe that the implication (iii) $\Rightarrow$ (i) in Theorem 0.J shows that (d) implies (b). Conversely, if (b) holds and W is a relatively compact open subset of Ω, then a positive superharmonic function v on Ω which satisfies $v \geq 1$ on $W \backslash K$ must satisfy the same inequality on the set

$$(W \backslash K) \cup \{X \in W \cap \partial K : \mathbf{R}^n \backslash K \text{ is not thin at } X\}$$

by fine continuity, and hence on the set

$$S = (W\backslash K) \cup \{X \in W \cap \partial K : \mathbf{R}^n\backslash K^\circ \text{ is not thin at } X\}.$$

Since the points of ∂K where $\mathbf{R}^n\backslash K^\circ$ is thin form a polar set (Theorem 0.E), it follows that $\widehat{R}_1^{W\backslash K} = \widehat{R}_1^S = \widehat{R}_1^{W\backslash K^\circ}$ (reductions are with respect to superharmonic functions on Ω), and this implies (d).

In preparation for the proof of Theorem 1.3 we give the following lemma which shows how far we can get by imitating the proof of Theorem 1.1.

Lemma 1.5. *Let K be a compact subset of $\mathbf{R}^n$ such that $\mathbf{R}^n\backslash K$ and $\mathbf{R}^n\backslash K^\circ$ are thin at the same points of K. Further, let $f \in C(\mathbf{R}^n)$, and let (W_m) be a decreasing sequence of open sets which satisfy*

$$K \subset W_m \subseteq \{X : \text{dist}(X, K) < 1/m\}.$$

(i) If $X_0 \in K^\circ$, then

$$H_f^{W_m}(X_0) \to H_f^{K^\circ}(X_0) \quad (m \to \infty).$$

(ii) If $X_0 \in \partial K$ and $\mathbf{R}^n\backslash K^\circ$ is not thin at X_0, then

$$H_f^{W_m}(X_0) \to f(X_0) \qquad (m \to \infty).$$

Proof. Let B be an open ball which contains $\overline{W_1}$. Reductions will be with respect to superharmonic functions on B. Let $\epsilon > 0$. Then there are positive continuous superharmonic functions u_1, u_2 on B such that

$$\left|f - (u_1 - u_2)\right| < \epsilon/3 \qquad \text{on } \overline{W_1}, \tag{1.5}$$

and hence

$$\left|H_f^{W_m} - H_{u_1}^{W_m} + H_{u_2}^{W_m}\right| < \epsilon/3 \qquad \text{on } W_m \qquad (m \geq 1). \tag{1.6}$$

We know that, if $X_0 \in K$, then

$$\begin{aligned} H_{u_k}^{W_m}(X_0) &= R_{u_k}^{B\backslash W_m}(X_0) = \widehat{R}_{u_k}^{B\backslash W_m}(X_0) \\ &\uparrow \widehat{R}_{u_k}^{B\backslash K}(X_0) \qquad (m \to \infty; k = 1, 2). \end{aligned} \tag{1.7}$$

Now, if v is a positive superharmonic function on B such that $v \geq u_k$ on $B\backslash K$, then fine continuity and the hypothesis concerning thinness ensure that $v \geq u_k$ on the set

$$S = (B\backslash K) \cup \{X \in \partial K : B\backslash K^\circ \text{ is not thin at } X\}.$$

Since the set of points of ∂K (which is a subset of $B\backslash K^\circ$) where $B\backslash K^\circ$ is thin forms a polar set, we obtain

$$\widehat{R}_{u_k}^{B\backslash K}(X_0) = \widehat{R}_{u_k}^{S}(X_0) = \widehat{R}_{u_k}^{B\backslash K^\circ}(X_0). \tag{1.8}$$

Combining (1.7) and (1.8), it follows that there exists m' (depending on X_0) such that

$$\widehat{R}_{u_k}^{B\backslash K^\circ}(X_0) - \epsilon/3 < H_{u_k}^{W_m}(X_0) \leq \widehat{R}_{u_k}^{B\backslash K^\circ}(X_0) \qquad (m \geq m'; k = 1,2). \tag{1.9}$$

We now deal separately with the cases numbered (i) and (ii) in the statement of the lemma. Firstly, if $X_0 \in K^\circ$, then

$$\widehat{R}_{u_k}^{B\backslash K^\circ}(X_0) = R_{u_k}^{B\backslash K^\circ}(X_0) = H_{u_k}^{K^\circ}(X_0) \qquad (k = 1,2). \tag{1.10}$$

Combining (1.6), (1.9) and (1.10), we obtain

$$\left| H_f^{W_m}(X_0) - H_{u_1}^{K^\circ}(X_0) + H_{u_2}^{K^\circ}(X_0) \right| < 2\epsilon/3 \qquad (m \geq m'). \tag{1.11}$$

Also, it follows from (1.5) that

$$\left| H_f^{K^\circ}(X_0) - H_{u_1}^{K^\circ}(X_0) + H_{u_2}^{K^\circ}(X_0) \right| < \epsilon/3.$$

This, together with (1.11) yields

$$\left| H_f^{W_m}(X_0) - H_f^{K^\circ}(X_0) \right| < \epsilon \qquad (m \geq m'),$$

as required.

Secondly, if $X_0 \in \partial K$ and $\mathbf{R}^n \backslash K^\circ$ is not thin at X_0, then

$$\widehat{R}_{u_k}^{B\backslash K^\circ}(X_0) = u_k(X_0). \qquad (k = 1,2).$$

This, together with (1.6) and (1.9), yields

$$\left| H_f^{W_m}(X_0) - u_1(X_0) + u_2(X_0) \right| < 2\epsilon/3 \qquad (m \geq m').$$

Combining this with (1.5) we again obtain the desired conclusion. This completes the proof of the lemma.

Thus, in particular, we have shown that, if $f \in C(\mathbf{R}^n) \cap \mathcal{H}(K^\circ)$ and if W_m and K are as in the lemma, then $H_f^{W_m}$ converges pointwise to f quasi-everywhere (i.e. except for a polar set) on K. In the next section we will establish uniform convergence on K. First we remark that one can find a sequence (W_m) of open sets as in the statement of the above lemma which are, in addition, regular for the Dirichlet problem. To see this, let

$$V_m = \{X \in \mathbf{R}^n : \operatorname{dist}(X, K) < 1/m\},$$

cover ∂V_m with a finite number of open balls B_k of radius $1/(2m)$ and define $W_m = V_m \backslash (\cup_k \overline{B}_k)$. There is then an exterior ball touching each boundary point of W_m, so W_m is regular.

1.4 Proof that (b) Implies (a) in Theorem 1.3

Suppose that K is a compact set such that $\mathbf{R}^n \backslash K$ and $\mathbf{R}^n \backslash K^\circ$ are thin at the same points, let $(W(m))$ be a decreasing sequence of regular open sets which satisfy $K \subset W(m) \subseteq \{X : \operatorname{dist}(X, K) < 1/m\}$, and let (L_k) be a sequence of compact sets such that $L_k \subset L_{k+1}^\circ$ for each k and such that $\cup_k L_k = K^\circ$. Also, let $\epsilon > 0$ and $u \in C(K) \cap \mathcal{H}(K^\circ)$. Tietze's extension theorem (see [Rud, §20.4]) asserts that there exists $\tilde{u}$ in $C(\mathbf{R}^n)$ such that $\tilde{u} = u$ on K.

It follows from the hypothesis on K that $\mathbf{R}^n \backslash \partial K$ is nowhere thin. Hence, by Theorem 1.1, there exists v_1 in $\mathcal{H}(\partial K)$ such that $|v_1 - \tilde{u}| < \epsilon/3$ on ∂K. In fact, this inequality must hold on an open set ω containing ∂K by continuity. We now choose k' and m_1 large enough to ensure that $\overline{W(m_1) \backslash L_{k'}^\circ} \subset \omega$.

It follows from Lemma 1.5 (case (i)) that $H_{\tilde{u}}^{W(m)} \to u$ locally uniformly on K°, so we can choose m_2 such that $m_2 \geq m_1$ and

$$\big|v(X) - u(X)\big| < \epsilon/3 \qquad (X \in L_{k'}), \tag{1.12}$$

where

$$v(X) = \begin{cases} H_{\tilde{u}}^{W(m_2)}(X) & (X \in W(m_2)) \\ \tilde{u}(X) & (X \in \partial W(m_2)). \end{cases}$$

Hence $|v - v_1| < 2\epsilon/3$ on $\partial L_{k'}$ and $|v - v_1| < \epsilon/3$ on $\partial W(m_2)$, and so $|v - v_1| < 2\epsilon/3$ on $W(m_2) \backslash L_{k'}$ by the maximum principle. Thus

$$|v - \tilde{u}| \leq |v - v_1| + |v_1 - \tilde{u}| < \epsilon \text{ on } W(m_2) \backslash L_{k'}. \tag{1.13}$$

Combining (1.12) and (1.13) we obtain $|v - u| < \epsilon$ on K, i.e. condition (a) of Theorem 1.3 holds.

1.5 Proof that (a) Implies (b) in Theorem 1.3

The implication "(a) $\Rightarrow$ (b)" of Theorem 1.3 is contained in the following result, stated here in a form which will be useful also in later chapters.

Theorem 1.6. *Let Ω be an open set in $\mathbf{R}^n$ and let E be a relatively closed subset of Ω. Suppose that, for each u in $C(E) \cap \mathcal{H}(E^\circ)$ and each positive number ϵ, there exists v in $\mathcal{H}(E)$ such that $|v - u| < \epsilon$ on E. Then $\mathbf{R}^n \backslash E$ and $\mathbf{R}^n \backslash E^\circ$ are thin at the same points of E.*

Proof. We dismiss the trivial case where $E = \Omega$. By deleting a suitable closed ball from $\Omega \backslash E$, if necessary, we can assume that Ω is a Greenian open set. Since $\mathbf{R}^n \backslash E \subseteq \mathbf{R}^n \backslash E^\circ$, it is enough to show that

$$[\mathbf{R}^n \backslash E^\circ \text{ is not thin at } Z] \Rightarrow [\mathbf{R}^n \backslash E \text{ is not thin at } Z] \qquad (Z \in E).$$

So suppose that $\mathbf{R}^n \backslash E^\circ$ is not thin at a fixed point Z of E, choose a positive number δ such that $\overline{B(Z,\delta)} \subset \Omega$ and let $\epsilon > 0$. Reductions below are with respect to superharmonic functions on Ω. Let $u^\#$ be the continuous potential on Ω described in Theorem 0.F. Since

$$\widehat{R}_{u^\#}^{[B(Z,\delta) \backslash B(Z,r)] \backslash E^\circ}(Z) \uparrow \widehat{R}_{u^\#}^{B(Z,\delta) \backslash E^\circ}(Z) = u^\#(Z) \qquad (r \downarrow 0),$$

there exists a positive number r_0 such that

$$\widehat{R}_{u^\#}^{[B(Z,\delta) \backslash B(Z,r_0)] \backslash E^\circ}(Z) \geq u^\#(Z) - \epsilon. \tag{1.14}$$

There is a measure μ with support in the compact set L defined by $L = [\overline{B(Z,\delta)} \backslash B(Z,r_0)] \backslash E^\circ$ for which the corresponding potential $G_\Omega \mu$ on Ω satisfies

$$G_\Omega \mu(X) = \widehat{R}_{u^\#}^{[B(Z,\delta) \backslash B(Z,r_0)] \backslash E^\circ}(X) \qquad (X \in \Omega). \tag{1.15}$$

Let $M = \max\{G_\Omega(Z,X) : X \in L\}$. There is a compact subset L' of L such that $\mu(L \backslash L') < \epsilon / M$, and such that the potential $u = G_\Omega(\mu|_{L'})$ is continuous on Ω, where $\mu|_{L'}$ denotes the restriction of μ to L' (see [Helm, 6.21]). Hence

$$G_\Omega \mu(Z) = \int_{L'} G_\Omega(Z,Y) d\mu(Y) + \int_{L \backslash L'} G_\Omega(Z,Y) d\mu(Y) < u(Z) + \epsilon,$$

and so, from (1.14) and (1.15),

$$u(Z) > u^\#(Z) - 2\epsilon. \tag{1.16}$$

Since $u \in C(E) \cap \mathcal{H}(E^\circ)$, we know by hypothesis that there exists v_ϵ in $\mathcal{H}(E)$ such that $|v_\epsilon - u| < \epsilon$ on E. In fact, by continuity, this inequality must hold on some open set which contains E. If we define

$$W_m = \{X \in B(Z,\delta) : \operatorname{dist}(X,E) < 1/m\} \qquad (m \in \mathbf{N}),$$

then, for large m,

$$\left|H_u^{W_m} - v_\epsilon\right| = \left|H_{u-v_\epsilon}^{W_m}\right| < \epsilon \text{ on } E \cap B(Z,\delta)$$

and thus

$$\left|\widehat{R}_u^{\Omega\setminus W_m} - u\right| \le \left|\widehat{R}_u^{\Omega\setminus W_m} - v_\epsilon\right| + |v_\epsilon - u| < 2\epsilon \text{ on } E \cap B(Z,\delta). \qquad (1.17)$$

Since $u^\# \ge G_\Omega\mu \ge u$ on Ω, we see from (1.16) and (1.17) that

$$\widehat{R}_{u^\#}^{\Omega\setminus[E\cap B(Z,\delta)]}(Z) > \widehat{R}_u^{\Omega\setminus[E\cap B(Z,\delta)]}(Z) > u(Z) - 2\epsilon > u^\#(Z) - 4\epsilon.$$

Since ϵ can be arbitrarily small, we conclude from Theorem 0.F that $B(Z,\delta)\setminus E$, and hence $\mathbf{R}^n\setminus E$, is not thin at Z. This completes the proof of Theorem 1.6, and thus also of Theorem 1.3.

1.6 Pole Pushing

The purpose of this section is to establish the following slight generalization of Walsh's Theorem (see §1.1). The argument is similar to that of Walsh [Wal, §7] which itself was inspired by Runge's work.

Theorem 1.7. *Let Ω be an open set in $\mathbf{R}^n$, let K be a compact subset of Ω, and suppose that $\Omega^*\setminus K$ is connected. Then, for each u in $\mathcal{H}(K)$ and each positive number ϵ, there exists v in $\mathcal{H}(\Omega)$ such that $|v - u| < \epsilon$ on K.*

The following lemma will first be proved.

Lemma 1.8. *Let K be a compact subset of $\mathbf{R}^n$, let $u \in \mathcal{H}(K)$ and $\epsilon > 0$. Then there are points $Y_1, Y_2, \ldots, Y_m$ in $\mathbf{R}^n\setminus K$ and constants $a_1, a_2, \ldots, a_m$ such that*

$$\left|u(X) - \sum_{k=1}^{m} a_k \phi_n\big(|X - Y_k|\big)\right| < \epsilon \qquad (X \in K).$$

Proof of Lemma. There is a bounded open set W containing K on which u is harmonic. Now let U be an open set such that $K \subset U$ and $\overline{U} \subset W$, and such that U is *admissible*. By this we mean that ∂U is a C^1 surface and that each point of ∂U is a one-sided boundary point of U. (This can achieved by constructing a C^1 function ψ on $\mathbf{R}^n$ valued 1 on K and 0 on $\mathbf{R}^n \backslash W$, and using Sard's Theorem [Ste, p.47] to choose a value of α in the interval $(0,1)$ such that the set $U = \{\psi > \alpha\}$ has the stated properties). Let σ denote surface area measure. If we apply Green's identity to $U \backslash \overline{B(X,\delta)}$ and let δ tend to 0, we obtain

$$u(X) = c_n \int_{\partial U} \left\{ \phi_n(|X-Y|) \frac{\partial u}{\partial n_Y}(Y) \right.$$
$$\left. -u(Y) \frac{\partial}{\partial n_Y} \phi_n(|X-Y|) \right\} d\sigma(Y) \quad (X \in K),$$

where n_Y denotes the outer unit normal to ∂U at Y and c_n is as defined in §0.2. The above integral can be approximated uniformly on K by a Riemann sum. Further, for a fixed choice of Y in ∂U, the derivative $X \mapsto (\partial/\partial n_Y)\phi_n(|X-Y|)$ can be uniformly approximated on K by a difference quotient. Hence the lemma is established.

Let $\mathcal{H}_k$ denote the vector space of all homogeneous harmonic polynomials of degree k on $\mathbf{R}^n$. (Thus $0 \in \mathcal{H}_k$.) If h is a harmonic function on the annulus $\{X : r < |X-Y| < R\}$, then we can write h as

$$h(X) = a + b\phi_n(|X-Y|) + \sum_{k=1}^{\infty} H_k(X-Y)$$
$$+ \sum_{k=1}^{\infty} |X-Y|^{2-n-2k} I_k(X-Y),$$

where a, b are constants, $H_k, I_k \in \mathcal{H}_k$ for each k, and the first (resp. second) series converges on $B(Y,R)$ (resp. on $\mathbf{R}^n \backslash \overline{B(Y,r)}$) and converges absolutely and uniformly on $B(Y, R-\epsilon)$ (resp. on $\mathbf{R}^n \backslash B(Y, r+\epsilon)$) for any positive number ϵ. (See [Bre2, Appendice] or [DuP, p.65]). We will refer to this as the *Laurent expansion* of h on the annulus. By suitably truncating the second series above we obtain the following result.

Lemma 1.9. *If h is harmonic on $\mathbf{R}^n \backslash \overline{B(Y,a)}$ and $b > a$ then, for each $\epsilon > 0$, there exists H in $\mathcal{H}(\mathbf{R}^n \backslash \{Y\})$ such that $|H - h| < \epsilon$ on $\mathbf{R}^n \backslash B(Y,b)$.*

Proof of Theorem 1.7. In view of Lemma 1.8, it remains to check that, if $Y_0 \in \mathbf{R}^n \backslash K$, then the function $h(X) = \phi_n(|X - Y_0|)$ can be uniformly approximated on K by functions in $\mathcal{H}(\Omega)$. First we consider the case where Y_0 belongs to a bounded component V of $\mathbf{R}^n \backslash K$. Since $\Omega^* \backslash K$ is connected, there exists a point Z in $V\backslash\Omega$. There are balls $\overline{B(Y_1, r_1)}, \ldots, \overline{B(Y_m, r_m)}$ in V, where $Y_m = Z$, such that $Y_{k-1} \in B(Y_k, r_k)$ for each k in $\{1, 2, \ldots, m\}$. By repeated application of Lemma 1.9 we obtain v in $\mathcal{H}(\mathbf{R}^n\backslash\{Z\})$ such that $|v - h| < \epsilon$ on K. Now consider the case where Y_0 belongs to the unbounded component V_∞ of $\mathbf{R}^n\backslash K$. There exists a sequence $\left(\overline{B(Y_k, r_k)}\right)$ of balls in V_∞ such that $r_k \leq 1$ for each k, such that $|Y_k| \to \infty$ as $k \to \infty$, and such that $Y_{k-1} \in B(Y_k, r_k)$ for each k. The result again follows from repeated application of Lemma 1.9, with ϵ replaced by $2^{-k}\epsilon$.

1.7 Runge Approximation

If Ω is an open set in $\mathbf{R}^n$, then a subset A of Ω is called Ω-*bounded* if $\overline{A}$ is a compact subset of Ω. If E is a relatively closed subset of Ω, then we denote by $\widehat{E}$ the union of E with all the Ω-bounded components of $\Omega\backslash E$. Thus $\widehat{E}$ depends not only on E but also on the choice of Ω. Below we present the harmonic analogue of Runge's Theorem (see §1.1). Its proof will be given §§1.8, 1.9.

Theorem 1.10. *Let Ω be an open set in $\mathbf{R}^n$ and K be a compact subset of Ω. The following are equivalent:*
(a) for each u in $\mathcal{H}(K)$ and each positive number ϵ there exists v in $\mathcal{H}(\Omega)$ such that $|v - u| < \epsilon$ on K;
(b) $\Omega\backslash\widehat{K}$ and $\Omega\backslash K$ are thin at the same points of K.

Unlike the holomorphic case (i.e. Runge's Theorem) we see that there are some compact sets with "holes" for which this type of approximation is possible. The following simple lemmas shed some light on what type of "holes" are permitted.

Lemma 1.11. *Suppose that $\Omega\backslash\widehat{K}$ and $\Omega\backslash K$ are thin at the same points of K. Then either $\widehat{K} = K$ or $\widehat{K}\backslash K$ is regular for the Dirichlet problem.*

Proof. Let $W = \widehat{K}\backslash K$, suppose that $W \neq \emptyset$ and that there exists an irregular boundary point Y of W. Then $\Omega\backslash W$ is thin at Y by Theorem 0.G, and so $\Omega\backslash\widehat{K}$ is thin at Y. By hypothesis $\Omega\backslash K$ is thin at Y. Hence W is thin at Y. This leads to the contradictory conclusion that Ω (being the union of W and $\Omega\backslash W$) is thin at Y, so W must be regular.

Lemma 1.12. *If $\Omega\backslash\widehat{K}$ and $\Omega\backslash K$ are thin at the same points of K, and $W = \widehat{K}\backslash K$, then $\partial W \subseteq \partial\widehat{K}$.*

Proof. Suppose that $W \neq \emptyset$ and let $A = \partial W \backslash \partial \widehat{K}$. Then $A \subseteq (\widehat{K})^\circ$, so $\Omega \backslash \widehat{K}$ is certainly thin at each point of A. By hypothesis, $\Omega \backslash K$ is thin at each point of A, so W is thin at each point of A. Thus A is a relatively open subset of ∂W which has zero harmonic measure for each component of W, by Theorem 0.H. Each point of A is therefore irregular for the Dirichlet problem on W, and it now follows from Lemma 1.11 that A is empty, as claimed.

Thus, for condition (b) of Theorem 1.10 to hold, it is necessary that the set $W = \widehat{K} \backslash K$ is regular and satisfies $\partial W \subseteq \partial \widehat{K}$. However, when $n \geq 3$, these conditions on W are not sufficient for (b) to hold, as the following example shows.

Example 1.13. Let $\Omega = \mathbf{R}^n$, where $n \geq 3$, let $\{Y'_k : k \in \mathbf{N}\}$ be a dense subset of $[0,1]^{n-1}$, and define

$$u(X') = \sum_{k=1}^{\infty} 2^{-k} \phi_{n-1}(|X' - Y'_k|) \qquad (X' \in \mathbf{R}^{n-1}),$$

$$K = \partial\left([0,1]^{n-1} \times [-1,0]\right) \cup \left\{(X', x_n) \in [0,1]^{n-1} \times (0,1] : u(X') \leq \phi_n(x_n)\right\},$$

and

$$F_y = \{X' \in \mathbf{R}^{n-1} : (X', y) \notin K\} \qquad (y > 0).$$

Then u is a superharmonic function on $\mathbf{R}^{n-1}$, and the lower semicontinuity of u ensures that K is closed and hence compact. Also, since

$$\{Y'_k : k \in \mathbf{N}\} \subseteq F_y \subseteq F_z \qquad (0 < y < z \leq 1),$$

it is clear that $\mathbf{R}^n \backslash K$ has exactly one bounded component, namely the cube $V = (0,1)^{n-1} \times (-1,0)$. (For future reference we note also that $K^\circ = \emptyset$.) Let $v(X', x_n) = u(X')$, so that v is superharmonic on $\mathbf{R}^n$. If $Z \in (0,1)^{n-1} \times \{0\}$, then

$$v(X', x_n) = u(X') > \phi_n(x_n) \geq \phi_n\Big(\big|(X', x_n) - Z\big|\Big) \qquad ((X', x_n) \in (0,1)^n \backslash K).$$

Since the Riesz measure associated with u does not charge $\{Z\}$, it follows (see Theorem 0.K) that $(0,1)^n \backslash K$, and hence $\mathbf{R}^n \backslash \widehat{K}$, is thin at Z. On the other hand, $\mathbf{R}^n \backslash K$ contains V and so is certainly not thin at Z. Thus condition (b) of Theorem 1.10 fails to hold, yet the only bounded component V of $\mathbf{R}^n \backslash K$ is regular for the Dirichlet problem and satisfies $\partial V \subseteq \partial \widehat{K}$.

1.8 Proof that (b) Implies (a) in Theorem 1.10

Let Ω be an open set in $\mathbf{R}^n$, let K be a compact subset of Ω, and suppose that $\Omega\backslash\widehat{K}$ and $\Omega\backslash K$ are thin at the same points of K. Also, let $u \in \mathcal{H}(K)$. Then u is harmonic on an open set ω which contains K. Since $\mathrm{dist}(K, \mathbf{R}^n\backslash\omega) > 0$, there are only finitely many Ω-bounded components $V_1, V_2, \ldots, V_m$ of $\Omega\backslash K$ which are not contained in ω. We assume that $m \neq 0$, for otherwise there is nothing to prove in view of Theorem 1.7.

Now fix k in $\{1, 2, \ldots, m\}$ and choose Y_k in V_k. There is a V_k-bounded open set U_k such that $Y_k \in U_k$ and $V_k\backslash U_k \subset \omega$. Let $G_k(\,\cdot\,,\,\cdot\,)$ denote the Green function for V_k. Lemma 1.11 shows that V_k is regular, and so $G_k(Y_k, \cdot)$ vanishes on ∂V_k. Now define

$$s_k(X) = a_k - b_k G_k(Y_k, X) \qquad (X \in V_k),$$

where

$$a_k = 1 + \sup\{u(X) : X \in \partial V_k\}$$

and b_k is a positive number chosen sufficiently large so that $s_k < u$ on ∂U_k. This ensures that the function defined by

$$u_k(X) = \begin{cases} s_k(X) & (X \in U_k) \\ \min\{s_k(X), u(X)\} & (X \in V_k\backslash U_k) \end{cases}$$

is superharmonic on $V_k\backslash\{Y_k\}$, and also that $u_k = u$ at all points of V_k which are sufficiently close to ∂V_k.

We repeat the above process for each k in $\{2, \ldots, m\}$ and define

$$h(X) = \begin{cases} u_k(X) & (X \in V_k; \quad k = 1, 2, \ldots, m) \\ u(X) & (X \in \omega\backslash \cup_k V_k). \end{cases}$$

It is now clear that h is superharmonic on $(\omega \cup \widehat{K})\backslash\{Y_1, Y_2, \ldots, Y_m\}$, and that the function

$$h(X) + \sum_{k=1}^{m} b_k \phi_n\left(|X - Y_k|\right)$$

has a superharmonic extension to $\omega \cup \widehat{K}$.

Let B be an open ball containing $\widehat{K}$, let G denote the Green function for B and, for each l in $\mathbf{N}$, let

$$W_l = \{X \in B : \mathrm{dist}(X, \widehat{K}) < 1/l\}.$$

Reductions below are with respect to superharmonic functions on B. If we fix k in $\{1, 2, \ldots, m\}$, then

$$\widehat{R}_{G(Y_k,\cdot)}^{B\backslash W_l}(X) \uparrow \widehat{R}_{G(Y_k,\cdot)}^{B\backslash \widehat{K}}(X) \qquad (X \in B; l \to \infty). \tag{1.18}$$

If we define

$$g_k(X) = G(Y_k, X) - \widehat{R}_{G(Y_k,\cdot)}^{B\backslash \widehat{K}}(X) \qquad (X \in B),$$

then g_k is a non-negative subharmonic function on $B\backslash\{Y_k\}$ which vanishes at points of ∂V_k where $\Omega\backslash\widehat{K}$ is not thin, and hence (by upper semicontinuity) has limit 0 at these points. The remaining points of ∂V_k are where $\Omega\backslash\widehat{K}$ is thin, and hence (by hypothesis) where $\Omega\backslash K$ is thin. Since $V_k \subset \Omega\backslash K$, these points form a set of zero harmonic measure for V_k (see Theorem 0.H). Clearly g_k is superharmonic on V_k, and the greatest harmonic minorant of g_k in V_k is a bounded harmonic function on V_k which vanishes at almost every (harmonic measure) boundary point. It is now easy to see that $g_k = G_k(Y_k, \cdot)$ on V_k.

Let $\epsilon > 0$. Since V_k is regular by Lemma 1.11, the set

$$L_k = \{X \in V_k : g_k(X) \geq \epsilon/(8mb_k)\}$$

is compact, and is a neighbourhood of Y_k. Further, $g_k \leq \epsilon/(8mb_k)$ on $V_k\backslash L_k^\circ$. By (1.18) and Dini's Theorem there exists l' such that

$$\begin{aligned} G(Y_k, X) - \widehat{R}_{G(Y_k,\cdot)}^{B\backslash W_{l'}}(X) &< g_k(X) + \frac{\epsilon}{8mb_k} \\ &\leq \frac{\epsilon}{4mb_k} \qquad (X \in \partial L_k; k = 1, \ldots, m), \end{aligned}$$

and so, by the maximum principle,

$$G(Y_k, X) - \widehat{R}_{G(Y_k,\cdot)}^{B\backslash W_{l'}}(X) < \frac{\epsilon}{4mb_k} \qquad (X \in K; k = 1, \ldots, m).$$

Thus, if we define

$$v_1(X) = h(X) + \sum_{k=1}^{m} b_k \left\{G(Y_k, X) - \widehat{R}_{G(Y_k,\cdot)}^{B\backslash W_{l'}}(X)\right\} + \epsilon/4,$$

we obtain a superharmonic function on $(\omega \cup \widehat{K}) \cap W_{l'}$ (see the conclusion of the third paragraph of this proof) which satisfies $h < v_1 < h + \epsilon/2$ on K. Hence

$$u < v_1 < u + \epsilon/2 \text{ on } K. \tag{1.19}$$

We can apply the above argument, replacing u by $-u$, to obtain a superharmonic function v_2 on an open set containing $\widehat{K}$ such that

$$-u < v_2 < -u + \epsilon/2 \text{ on } K. \tag{1.20}$$

It follows that $-v_2 < u < v_1$ on K. Hence $-v_2 < v_1$ on an open set which contains K, and thus also (by the maximum principle) on an open set W which contains $\widehat{K}$. The greatest harmonic minorant w of v_1 on W therefore satisfies $-v_2 \leq w \leq v_1$ on $\widehat{K}$. From (1.19) and (1.20) it follows that $|w - u| < \epsilon/2$ on K.

Since $w \in \mathcal{H}(\widehat{K})$ and $\Omega^* \backslash \widehat{K}$ is connected (by the definition of $\widehat{K}$) we can apply Theorem 1.7 to obtain v in $\mathcal{H}(\Omega)$ such that $|v - w| < \epsilon/2$ on $\widehat{K}$. Thus $|v - u| < \epsilon$ on K, and the proof is complete.

1.9. Proof that (a) Implies (b) in Theorem 1.10

The implication "(a) $\Rightarrow$ (b)" of Theorem 1.10 is contained in the following result.

Theorem 1.14. *Let Ω be an open set in $\mathbf{R}^n$ and E be a relatively closed subset of Ω. Suppose that, for each compact subset L of Ω, there is a compact subset of Ω which contains every Ω-bounded component of $\Omega \backslash E$ that intersects L. Suppose further that, for each u in $\mathcal{H}(E)$ and each positive number ϵ, there exists v in $\mathcal{H}(\widehat{E})$ such that $|v - u| < \epsilon$ on E. Then $\Omega \backslash \widehat{E}$ and $\Omega \backslash E$ are thin at the same points of E.*

Proof. We may suppose that $E \neq \Omega$. If $\widehat{E} = \Omega$, then we choose Y in $\widehat{E} \backslash E$ and let $u(X) = \phi_n(|X - Y|)$. By hypothesis and the minimum principle there exists v in $\mathcal{H}(\Omega)$ such that $u - v + \epsilon$ is a non-constant positive superharmonic function on Ω, whence Ω is Greenian. If $\widehat{E} \neq \Omega$ then, by deleting a suitable closed ball from $\Omega \backslash \widehat{E}$ when $n = 2$, we may again assume that Ω is Greenian.

Since $\Omega \backslash \widehat{E} \subseteq \Omega \backslash E$, it is enough to show that

$$[\Omega \backslash E \text{ is not thin at } Z] \Rightarrow [\Omega \backslash \widehat{E} \text{ is not thin at } Z] \qquad (Z \in E).$$

So suppose that $\Omega \backslash E$ is not thin at a fixed point Z of E, choose a positive number δ such that $\overline{B(Z, \delta)} \subset \Omega$ and let $\epsilon > 0$. Reductions below are with respect to superharmonic functions on Ω. Let $u^\#$ be the continuous potential on Ω described in Theorem 0.F, and let L be a compact subset of $B(Z, \delta) \backslash E$ such that

$$\widehat{R}^L_{u^\#}(Z) > \widehat{R}^{B(Z,\delta) \backslash E}_{u^\#}(Z) - \epsilon.$$

Thus

$$u(Z) > u^\#(Z) - \epsilon, \tag{1.21}$$

where

$$u(X) = \widehat{R}^L_{u^\#}(X) \qquad (X \in \Omega).$$

The function u is a potential on Ω with corresponding measure supported by the set L, and so $u \in \mathcal{H}(E)$. We now adopt an argument similar to the proof of Theorem 1.6. For each positive number ϵ there exists, by hypothesis, v_ϵ in $\mathcal{H}(\widehat{E})$ such that $|v_\epsilon - u| < \epsilon$ on E. By continuity this inequality remains true on some open set containing E and hence $\partial\widehat{E} \cap \Omega$. We define W to be the union of $B(Z,\delta)$ with all Ω-bounded components of $\Omega\backslash E$ which intersect $B(Z,\delta)$. Then W is Ω-bounded by hypothesis, and $\partial W \subseteq E \cup \partial B(Z,\delta)$. Thus, if we define

$$W_m = \{X \in W : \operatorname{dist}(X, \widehat{E}) < 1/m\} \qquad (m \in \mathbf{N}),$$

then, for large values of m,

$$\left|R_u^{\Omega\backslash W_m} - v_\epsilon\right| = \left|H_u^{W_m} - v_\epsilon\right| = \left|H_{u-v_\epsilon}^{W_m}\right| < \epsilon \text{ on } E \cap B(Z,\delta),$$

and so

$$\left|\widehat{R}_u^{\Omega\backslash W_m} - u\right| \le \left|\widehat{R}_u^{\Omega\backslash W_m} - v_\epsilon\right| + |v_\epsilon - u| < 2\epsilon \text{ on } E \cap B(Z,\delta). \qquad (1.22)$$

Since $u^\# \ge u$ on Ω, we see from (1.21) and (1.22) that

$$\widehat{R}_{u^\#}^{\Omega\backslash[\widehat{E}\cap W]}(Z) > \widehat{R}_u^{\Omega\backslash[\widehat{E}\cap W]}(Z) > u(Z) - 2\epsilon > u^\#(Z) - 3\epsilon.$$

Since ϵ can be arbitrarily small, we conclude from Theorem 0.F that $W\backslash\widehat{E}$, and hence $\Omega\backslash\widehat{E}$, is not thin at Z. This completes the proof of Theorem 1.14, and thus also of Theorem 1.10.

1.10 An Analogue of Mergelyan's Theorem

Theorems 1.3 and 1.10, when combined, immediately yield the following analogue of Mergelyan's Theorem (see §1.1).

Theorem 1.15. *Let Ω be an open set in $\mathbf{R}^n$ and K be a compact subset of Ω. The following are equivalent:*
(a) for each u in $C(K) \cap \mathcal{H}(K^\circ)$ and each positive number ϵ, there exists v in $\mathcal{H}(\Omega)$ such that $|v - u| < \epsilon$ on K;
(b) $\Omega\backslash\widehat{K}$ and $\Omega\backslash K^\circ$ are thin at the same points of K.

In fact, there is a more direct proof of this result which we indicate below.

Proof that (b) implies (a). Suppose that (b) holds, let $u \in C(K) \cap \mathcal{H}(K^\circ)$, and let $\epsilon > 0$. Since $\Omega\backslash\widehat{K} \subseteq \Omega\backslash K \subseteq \Omega\backslash K^\circ$, the sets $\Omega\backslash\widehat{K}$ and $\Omega\backslash K$

are thin at the same points of K. It follows from Lemmas 1.11 and 1.12 that the set $W = \widehat{K}\backslash K$ is either empty, or is regular for the Dirichlet problem and satisfies $\partial W \subseteq \partial\widehat{K}$. Hence, by solving the Dirichlet problem in W, we can extend u to a function $\tilde{u}$ in $C(\widehat{K}) \cap \mathcal{H}((\widehat{K})^\circ)$.

Next, since $\Omega\backslash\widehat{K} \subseteq \Omega\backslash(\widehat{K})^\circ \subseteq \Omega\backslash K^\circ$, the sets $\Omega\backslash\widehat{K}$ and $\Omega\backslash(\widehat{K})^\circ$ are thin at the same points of K, and hence at the same points of $\widehat{K}$. It follows from Theorem 1.3 that there exists h in $\mathcal{H}(\widehat{K})$ such that $|h - \tilde{u}| < \epsilon/2$ on $\widehat{K}$. Also, from Theorem 1.7, there exists v in $\mathcal{H}(\Omega)$ such that $|v - h| < \epsilon/2$ on $\widehat{K}$, since $\Omega^*\backslash\widehat{K}$ is certainly connected. Hence $|v - u| < \epsilon$ on K, as required.

Proof that (a) implies (b). This requires only a minor modification to the proof of Theorem 1.6. The details are left to the reader.

1.11 The Case where $n = 2$

Let $n \geq 3$, let K be as in Example 1.13 and let $K_1 = \widehat{K}$. Then $\mathbf{R}^n\backslash K_1$ is thin at each point of $(0,1)^{n-1} \times \{0\}$, whereas $\mathbf{R}^n\backslash K_1^\circ$ is not (since $K^\circ = \emptyset$). Thus K_1 satisfies condition (b) of Theorem 1.10 (because $\widehat{K}_1 = K_1$), but not condition (b) of Theorem 1.15. As the next result shows, this distinction between the two types of approximation does not apply when $n = 2$. In this respect harmonic approximation in $\mathbf{R}^2$ is analogous to holomorphic approximation in $\mathbf{C}$: compare Runge's Theorem and Mergelyan's Theorem in §1.1.

Corollary 1.16 *Let Ω be an open set in $\mathbf{R}^2$ and K be a compact subset of Ω. The following are equivalent:*
(a) for each u in $\mathcal{H}(K)$ and each positive number ϵ, there exists v in $\mathcal{H}(\Omega)$ such that $|v - u| < \epsilon$ on K;
(b) for each u in $C(K) \cap \mathcal{H}(K^\circ)$ and each positive number ϵ, there exists v in $\mathcal{H}(\Omega)$ such that $|v - u| < \epsilon$ on K;
(c) $\partial\widehat{K} = \partial K$.

We observe below that Corollary 1.16 follows from the earlier results of this chapter without claiming that this is the most direct way of establishing it.

Proof. Clearly (b) implies (a). Also, Theorem 1.14 and Lemma 1.12 together show that (a) implies (c). Now suppose that (c) holds and let Y be a point of K at which $\Omega\backslash\widehat{K}$ is thin. Then, since $n = 2$, there are arbitrarily small circles, centred at Y, which are contained in $\widehat{K}$ (see Theorem 0.D). Hence $Y \in (\widehat{K})^\circ$. By hypothesis, it follows that $Y \notin \partial K$, so $Y \in K^\circ$, and thus $\Omega\backslash K^\circ$ is certainly thin at Y. It has now been shown that $\Omega\backslash\widehat{K}$ and $\Omega\backslash K^\circ$ are thin at the same points of K, so (b) follows from Theorem 1.15.

Notes

Theorem 1.1 is due to Keldyš [Kel] (1941), Brelot [Brel] (1945) and Deny [Den1] (1945), while Theorem 1.3 is due to Keldyš [Kel] (1941) and Deny [Den2] (1949). (Condition (d) in §1.3 was pointed out by Labrèche [Lab].)

Deny's proof [Den2] of Theorem 1.3 is based on methods of functional analysis, but it is not as short as the brevity of the explanation given there suggests. The main thrust of this book is approximation on sets which need not be compact, and here the functional analysis approach appears to be less useful. The arguments given in this chapter will be adapted later in the book to deal with approximation on non-compact sets.

In connection with Theorem 1.3 we mention the book by Landkof [Lan, Chapter V §5], including the historical notes given there, and also a recent interesting survey article by Hedberg [Hed]. The climax of that paper is [Hed, Theorem 11.9], which adds some further equivalent conditions to the theorem. We also mention Debiard and Gaveau [DG], who characterize, in terms of "fine harmonicity", the functions on a compact set K which can be approximated uniformly by members of $\mathcal{H}(K)$.

Theorems 1.10 and 1.15 are special cases of theorems in [Gar3] concerning approximation on relatively closed subsets of Ω. They can also be derived from results of Bliedtner and Hansen [BH] concerning superharmonic approximation on compact sets, which were proved in an abstract setting. When $\Omega = \mathbf{R}^2$, the implication "(c) $\Rightarrow$ (b)" of Corollary 1.16 follows from a result of Walsh and Lebesgue [Wal, p.503]. The implication "(b) $\Rightarrow$ (c)" of the same result is contained in [GGO1, Theorem 2]. We remark that a weak form of Corollary 1.16 has been used to give an alternative proof of Mergelyan's Theorem: see [CarL].

Various generalizations of results in this chapter to more abstract settings can be found in [Pra1], [Pra2], [EdVS], [VS], [EK], [HV], [BH] and [GGG1].

2 Fusion of Harmonic Functions

2.1 Introduction

The material of this chapter has its roots in the following result of Alice Roth [Rot3] concerning "fusion" of rational functions.

Roth's Fusion Lemma. *For each pair K_1, K_2 of disjoint compact subsets of $\mathbf{C}^*$, there is a positive constant C with the following property: if r_1, r_2 are rational functions satisfying $\left|r_1(z) - r_2(z)\right| < \epsilon$ on some compact subset K of $\mathbf{C}^*$, then there is a rational function r such that*

$$\left|r(z) - r_k(z)\right| < C\epsilon \qquad (z \in K_k \cup K; k = 1, 2).$$

Gauthier, Goldstein and Ow subsequently developed corresponding fusion lemmas for harmonic functions in $\mathbf{R}^2$ with logarithmic singularities [GGO1], and later for harmonic functions in $\mathbf{R}^n (n \geq 3)$ with Newtonian singularities [GGO2]. Further progress was made by Gauthier and Hengartner [GH]. The significance of these contributions can be seen from the observation that most of the remaining material in these lecture notes depends on some adaptation of their work. In this chapter we establish one such fusion result, due to Armitage and Goldstein [AG3], which will be useful in subsequent chapters. Later, in Chapter 5, this fusion result will reappear in a modified form.

Throughout this chapter Ω will denote a connected open set in $\mathbf{R}^n$ with Green function $G(\,\cdot\,,\,\cdot\,)$. We fix a point Q in Ω and define $g(\,\cdot\,) = \min\{1, G(Q,\,\cdot\,)\}$. The fusion result that we shall establish achieves better-than-uniform approximation: the error of approximation is at most a multiple of $g(\,\cdot\,)$. Since g continuously vanishes at all regular boundary points of Ω, the approximation is very good near most of the boundary. This will be of particular significance for the application to the Dirichlet problem given in Chapter 7.

2.2 Preliminary Lemmas

We continue to write $\partial^*\Omega$ for the boundary of Ω in the compactified space $\mathbf{R}^n \cup \{\infty\}$, and recall that, if Ω is unbounded and $n \geq 3$, then ∞ is a regular boundary point for the Dirichlet problem on Ω. We begin with two elementary lemmas in which g appears as a majorizing function.

Lemma 2.1. *Let K be compact subset of Ω, and let s be a subharmonic function on $\Omega\backslash K$ which is bounded above. If*

$$\limsup_{X\to Y}\{s(X)/g(X)\} \leq a \qquad (Y \in \partial K),$$

where $a \geq 0$, and

$$\limsup_{X\to Y} s(X) \leq 0$$

for every regular point Y of $\partial^*\Omega$, then $s \leq ag$ on $\Omega\backslash K$.

Proof. Let u^* be the upper semicontinuous regularization of the function

$$u(X) = \begin{cases} [s(X) - ag(X)]^+ & (X \in \Omega\backslash K) \\ 0 & (X \in (\mathbf{R}^n\backslash\Omega) \cup K). \end{cases}$$

Then u^* is subharmonic on $\mathbf{R}^n$. (See, for example, [Doo, 1.V.5].) Further, u^* is bounded (resp. $u^*(X) \to 0$ as $|X| \to \infty$) if $n = 2$ (resp. $n \geq 3$). In either case u^* is constant with value 0 on $\mathbf{R}^n$, as required.

We write $c(a, b, \ldots)$ for a positive constant, depending at most on $a, b, \ldots$, not necessarily the same on any two occurrences. We also write ∇_j for the gradient operator with respect to the jth argument $(j = 1, 2)$ of a function of two vector variables.

Lemma 2.2. *Let $K \subset \omega$ and $\overline{\omega} \subset \Omega$, where K is compact and ω is open. Then*

$$G(X,Y) + |\nabla_2 G(X,Y)| \leq c(K,\omega,\Omega,Q)g(X) \qquad (X \in \Omega\backslash\omega; Y \in K).$$

Proof. Let $a = \operatorname{dist}(K, \mathbf{R}^n\backslash\omega)$. It follows from Harnack's inequalities (see [Bre2, Appendice IX]) that

$$|\nabla_2 G(X,Y)| \leq (n/a)G(X,Y) \qquad (X \in \Omega\backslash\omega; Y \in K). \tag{2.1}$$

Next, if we fix P in K, Harnack's inequalities yield

$$G(X,Y) \leq c(K,\omega)G(X,P) \qquad (X \in \Omega\backslash\omega; Y \in K). \tag{2.2}$$

Finally, if we define

$$d = \inf\{g(X)/G(X,P) : X \in \partial B(P,a)\},$$

then

$$G(X,P) \leq d^{-1}g(X) \qquad (X \in \Omega\backslash\omega). \tag{2.3}$$

The lemma follows, on combining (2.1)–(2.3).

If $A \subseteq \mathbf{R}^n$, then we use $\mathcal{I}(A)$ to denote the collection of functions which are harmonic, apart from isolated singularities, on an open set containing A.

Now let U be an Ω-bounded admissible (see §1.6) open set. Further, let $\psi \in C^1(\overline{U}) \cap C^2(U)$ and define

$$w(X) = c_n \int_{\partial U} \left\{ \psi(Y) \frac{\partial}{\partial n_Y} G(X,Y) - G(X,Y) \frac{\partial \psi}{\partial n_Y}(Y) \right\} d\sigma(Y) \qquad (X \in \Omega_0),$$

where $\Omega_0 = \Omega\backslash(\partial U \cap \operatorname{supp}\psi)$, and where $\operatorname{supp}\psi$ denotes the compact support of the function ψ. (Recall that $c_n^{-1} = \sigma_n \max\{n-2, 1\}$ and that n_Y denotes the outer unit normal to ∂U at Y.) The function w arises in Lemma 2.4 below from the use of Green's identity. Differentiation under the integral sign shows that w is harmonic on Ω_0. In the next result we approximate w by harmonic functions with isolated singularities in the set $\partial U \cap \operatorname{supp}\psi$.

Lemma 2.3. *Let U, ψ and w be as above, let ω be an Ω-bounded open set which contains $\partial U \cap \operatorname{supp}\psi$, and let $\epsilon > 0$. Then there exists w_1 in $\mathcal{I}(\Omega) \cap \mathcal{H}(\Omega_0)$ such that*

$$|w_1(X) - w(X)| \leq \epsilon g(X) \qquad (X \in \Omega\backslash\omega). \tag{2.4}$$

Proof. Since g has a positive lower bound on the set $\partial\omega$, we can choose points $Y_1, Y_2, \ldots, Y_m$ in $\partial U \cap \operatorname{supp}\psi$ such that

$$|w_1(X) - w(X)| \leq \epsilon g(X) \qquad (X \in \partial\omega), \tag{2.5}$$

where w_1 is a Riemann sum of the form

$$w_1(X) = c_n \sum_{k=1}^{m} a_k \left\{ \psi(Y_k) \frac{\partial}{\partial n_{Y_k}} G(X,Y_k) - G(X,Y_k) \frac{\partial \psi}{\partial n_{Y_k}}(Y_k) \right\},$$

where $a_k > 0$ for each k, and where $\sum_k a_k = \sigma(\partial U \cap \operatorname{supp}\psi)$. It follows from Lemma 2.2 that

$$|w(X)| \leq c(\partial U \cap \operatorname{supp} \psi, \omega, \Omega, Q) c(\psi) g(X) \qquad (X \in \Omega \backslash \omega),$$

and a similar inequality holds for w_1. Hence

$$|w_1(X) - w(X)| \leq c(U, \omega, \Omega, Q, \psi) g(X) \qquad (X \in \Omega \backslash \omega). \tag{2.6}$$

The inequality (2.4) now follows from (2.5), (2.6) and Lemma 2.1.

We will use λ to denote n-dimensional Lebesgue measure.

Lemma 2.4. *Let U, ψ and w be as above. Then*

$$w(X) + c_n \int_U G(X,Y) \Delta\psi(Y)\, d\lambda(Y) = \begin{cases} -\psi(X) & (X \in U) \\ 0 & (X \in (\Omega \backslash \overline{U}) \cup (\partial U \backslash \operatorname{supp} \psi)). \end{cases} \tag{2.7}$$

Proof. If $X \in \Omega \backslash \overline{U}$, then (2.7) follows by applying Green's (second) identity to U. If $X \in U$, then (2.7) follows by applying Green's identity in $U \backslash \overline{B(X, \delta)}$ and letting δ tend to 0. If $X \in \partial U \backslash \operatorname{supp} \psi$, then we choose δ such that $\overline{B(X, \delta)}$ does not intersect $\operatorname{supp} \psi$, and let U' be an admissible open subset of U such that $X \notin \overline{U'}$ and $U' \backslash B(X, \delta) = U \backslash B(X, \delta)$. The values of the integral which defines w and the integral in (2.7) are unchanged if they are taken over $\partial U'$ and U' respectively. Since $X \in \Omega \backslash \overline{U'}$, (2.7) follows from Green's identity as before.

The following refinement of Lemma 1.9 will be useful below and also in Chapter 5.

Lemma 2.5. *If h is harmonic on $\mathbf{R}^n \backslash \overline{B(Y, a)}$ and $b > a$ then, for each choice of positive numbers ϵ and d, there exists H in $\mathcal{H}(\mathbf{R}^n \backslash \{Y\})$ such that*

$$|H(X) - h(X)| < \epsilon (1 + |X|)^{-d} \qquad (X \in \mathbf{R}^n \backslash B(Y, b)).$$

Proof. Without loss of generality we may assume that $Y = O$. It follows from the Laurent expansion (see the discussion preceding Lemma 1.9) that we can write h as

$$h(X) = h_1(X) + c\phi_n(|X|) + \sum_{k=1}^{\infty} |X|^{2-n-2k} H_k(X) \quad (X \in \mathbf{R}^n \backslash \overline{B(O, a)}), \tag{2.8}$$

where c is a constant, $h_1 \in \mathcal{H}(\mathbf{R}^n)$ and $H_k \in \mathcal{H}_k$ for each k, and the series is absolutely and uniformly convergent on $\partial B(O,b)$. Thus there exists a positive integer l such that $l > d+1-n$ and

$$\sum_{k=l+1}^{\infty} |H_k(X)| b^{2-n-2k} < \epsilon(1+b)^{-d} \quad (X \in \partial B(O,b)). \tag{2.9}$$

We now define

$$H(X) = h_1(X) + c\phi_n(|X|) + \sum_{k=1}^{l} |X|^{2-n-2k} H_k(X) \quad (X \neq O), \tag{2.10}$$

so that $H \in \mathcal{H}(\mathbf{R}^n \backslash \{O\})$. Defining $\tilde{X} = b|X|^{-1}X$, it follows from (2.8)–(2.10) and the homogeneity of H_k that

$$\begin{aligned}
|H(X) - h(X)| &\leq \sum_{k=l+1}^{\infty} \big(|X|/b\big)^{2-n-k} |H_k(\tilde{X})| b^{2-n-2k} \\
&\leq \big(|X|/b\big)^{1-n-l} \sum_{k=l+1}^{\infty} |H_k(\tilde{X})| b^{2-n-2k} \qquad (|X| \geq b) \\
&\leq (|X|/b)^{-d} \epsilon (1+b)^{-d} \\
&\leq \epsilon \big(1+|X|\big)^{-d},
\end{aligned}$$

as required.

Now we modify Lemma 2.5 to deal with approximation by harmonic functions on $\Omega \backslash \{Y\}$, where the error of approximation involves the function g.

Lemma 2.6. *Let $0 < a < b$, and suppose that $\overline{B(Y,b)} \subset \Omega$. If h is harmonic on $\Omega \backslash \overline{B(Y,a)}$ and $\epsilon > 0$, then there exists H in $\mathcal{H}\big(\Omega \backslash \{Y\}\big)$ such that*

$$\big|H(X) - h(X)\big| < \epsilon g(X) \qquad \big(X \in \Omega \backslash B(Y,b)\big).$$

Proof. By considering the Laurent expansion of h in $B(Y,b) \backslash \overline{B(Y,a)}$, we can write $h = h_1 + h_2$, where $h_1 \in \mathcal{H}\big(\mathbf{R}^n \backslash \overline{B(Y,a)}\big)$ and $h_2 \in \mathcal{H}(\Omega)$. Let $\epsilon > 0$ and define

$$\epsilon' = (\epsilon/2) \inf\big\{g(X) : X \in \partial B(Y,b)\big\}.$$

It follows from Lemma 2.5 that there exists u_1 in $\mathcal{H}\big(\mathbf{R}^n \backslash \{Y\}\big)$ such that

$$|u_1 - h_1| < \epsilon' \big(1+|X|\big)^{-1} \qquad \big(X \in \mathbf{R}^n \backslash B(Y,b)\big).$$

We define

$$u_2 = H^{\Omega}_{u_1 - h_1} \quad \text{on } \Omega.$$

(If Ω is unbounded, we assign the value 0 to $u_1 - h_1$ at ∞.) Then u_2 is bounded and

$$|u_1 - u_2 - h_1| \le |u_1 - h_1| + |u_2| < 2\epsilon' \le \epsilon g \quad \text{on } \partial B(Y, b).$$

It follows from Lemma 2.1 that, if we define $H = h_2 + u_1 - u_2$ on $\Omega\backslash\{Y\}$, then

$$|H - h| = |u_1 - u_2 - h_1| < \epsilon g \quad \text{on } \Omega\backslash B(Y, b),$$

as required.

2.3 A Fusion Result

The main result of the chapter is as follows.

Theorem 2.7. *Let K, E_1 be compact subsets of Ω and let E_2 be a relatively closed subset of Ω which is disjoint from E_1. Then there is a positive constant C with the following property: if u_1, $u_2 \in \mathcal{I}(\Omega) \cap \mathcal{H}(K)$ and $|u_1 - u_2| < \epsilon$ on K, then there exists u in $\mathcal{I}(\Omega)$ such that*

$$\left|(u - u_k)(X)\right| < C\epsilon g(X) \qquad (X \in K \cup E_k; k = 1, 2). \tag{2.11}$$

Inequalities such as (2.11) are to be interpreted in the sense that $u - u_k$ has, at worst, removable singularities in $K \cup E_k$, and the inequality is valid when this function is suitably redefined at these points. Before proving Theorem 2.7 we make the following two observations.

Firstly, it is enough to consider the case where $u_2 \equiv 0$. For, if this case is established, then in the general case we can find v in $\mathcal{I}(\Omega)$ such that

$$\left|\left(v - (u_1 - u_2)\right)(X)\right| < C\epsilon g(X) \qquad (X \in K \cup E_1)$$

and

$$\left|v(X)\right| < C\epsilon g(X) \qquad (X \in K \cup E_2).$$

Thus (2.11) is satisfied by defining $u = v + u_2$.

Secondly, it is sufficient to consider the case where $K \neq \emptyset$. For, if this case is proved, then we can deal with the case where $K = \emptyset$ as follows. Let K' be some closed ball in $\Omega\backslash(\widehat{E}_1 \cup E_2)$, let C be the positive constant in the statement of the theorem for the sets K', E_1, E_2, and let

$$\delta = (\epsilon/2)\inf\Big(\big\{Cg(X) : X \in \widehat{E}_1\big\} \cup \{1\}\Big).$$

Let $X_1, X_2, \ldots, X_m$ denote the singularities of u_1 in $\widehat{E}_1$. Then, by considering the appropriate Laurent expansions of u_1, we can write u_1 as $u_1 = f_0 + f_1 + \cdots + f_m$, where $f_0 \in \mathcal{I}(\Omega) \cap \mathcal{H}(\widehat{E}_1)$ and $f_k \in \mathcal{H}(\Omega \backslash \{X_k\})$ for each k in $\{1, 2, \ldots, m\}$. It follows from Theorem 1.7 that there is a harmonic function H on Ω such that $|H - f_0| < \delta$ on $\widehat{E}_1$ and $\big|H + (f_1 + \ldots + f_m)\big| < \delta$ on K'. Thus the function u_1', defined by $u_1' = H + f_1 + \ldots + f_m$, belongs to $\mathcal{I}(\Omega) \cap \mathcal{H}(K')$ and satisfies

$$\big|(u_1' - u_1)(X)\big| < C\epsilon g(X)/2 \qquad (X \in E_1) \tag{2.12}$$

and

$$\big|u_1'(X)\big| < \epsilon/2 \qquad (X \in K'). \tag{2.13}$$

It now follows from (2.13) and the assumed case that there exists u in $\mathcal{I}(\Omega)$ such that

$$\big|(u - u_1')(X)\big| < C\epsilon g(X)/2 \qquad (X \in E_1) \tag{2.14}$$

and

$$\big|u(X)\big| < C\epsilon g(X)/2 \qquad (X \in E_2).$$

Combining (2.12) and (2.14) we obtain

$$\big|(u - u_1)(X)\big| < C\epsilon g(X) \qquad (X \in E_1).$$

The last two inequalities establish (2.11) in the case where $u_2 \equiv 0$.

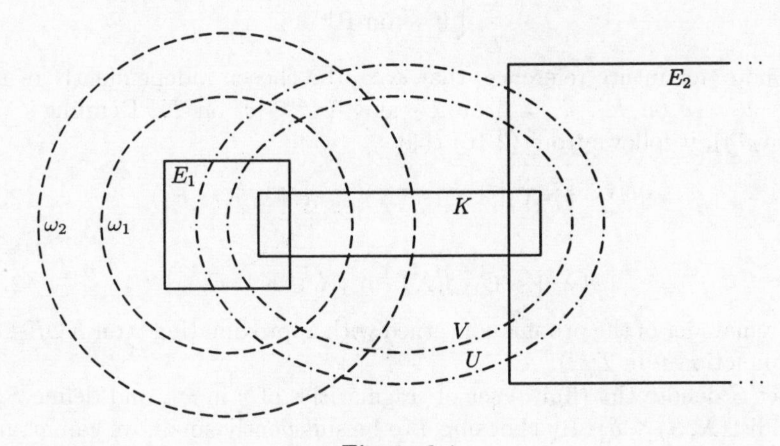

Figure 2.1

Proof of Theorem 2.7. In view of the preceding remarks we will now prove Theorem 2.7 under the additional assumptions that $u_2 \equiv 0$ and $K \neq \emptyset$. Let

ω_1, ω_2 be Ω-bounded admissible open sets such that $E_1 \subset \omega_1$, $\overline{\omega}_1 \subset \omega_2$ and $\overline{\omega}_2 \subset \Omega \backslash E_2$. Also, let U be an Ω-bounded open set which contains K. Since $u_1 \in \mathcal{H}(K)$ and $|u_1| < \epsilon$ on K, we can find an open set V such that $K \subset V \subseteq U$ and $\sup_V |u_1| < \epsilon$ (see Figure 2.1). Further, by allowing some flexibility in the initial choice of ω_1, it can be arranged that $\omega_1 \cup V$ is admissible, and it can also be arranged that $u_1 \in \mathcal{H}(\overline{V})$.

Next we deal with the possibility that u_1 may have singularities on the compact set $\partial\omega_1 \cup \partial\omega_2$. If we denote these singularities by $X_1, X_2, \ldots, X_m$, then we can write $u_1 = f_0 + f_1 + \cdots + f_m$, where $f_0 \in \mathcal{I}(\Omega) \cap \mathcal{H}(\partial\omega_1 \cup \partial\omega_2)$ and $f_k \in \mathcal{H}(\mathbf{R}^n \backslash \{X_k\})$ for each k in $\{1, 2, \ldots, m\}$. Since $X_k \notin E_1 \cup E_2$ (by our choice of ω_1 and ω_2) and $X_k \notin \overline{V}$, we can choose an open ball B_k whose centre Y_k is outside the set $\partial\omega_1 \cup \partial\omega_2$ such that $X_k \in B_k \subset \Omega \backslash (E_1 \cup E_2 \cup \overline{V})$. It follows from Lemma 2.6 that there exists h_k in $\mathcal{H}(\Omega \backslash \{Y_k\})$ such that $|h_k - f_k| < \epsilon g/m$ on $\Omega \backslash B_k$. If we define $v = f_0 + h_1 + \cdots + h_m$, then $v \in \mathcal{I}(\Omega) \cap \mathcal{H}(\partial\omega_1 \cup \partial\omega_2 \cup \overline{V})$ and

$$\big|(v - u_1)(X)\big| < \epsilon g(X) \qquad (X \in E_1 \cup E_2 \cup \overline{V}). \tag{2.15}$$

In particular, it follows that

$$\big|v(X)\big| < 2\epsilon \qquad (X \in \overline{V}). \tag{2.16}$$

We now choose ϕ in $C^\infty(\mathbf{R}^n)$ such that $0 \le \phi \le 1$ on $\mathbf{R}^n$, $\phi = 1$ on ω_1 and $\phi = 0$ on $\mathbf{R}^n \backslash \omega_2$, and define

$$\psi = \begin{cases} \phi v & \text{on } \omega_2 \\ 0 & \text{on } \mathbf{R}^n \backslash \omega_2. \end{cases}$$

(We note, for future reference, that ϕ can be chosen independently of u_1.) Thus $\psi = v$ on E_1, $\psi = 0$ on E_2 and $|\psi| \le |v|$ on K. Defining $C_1 = 2/(\inf_K g)$, it follows from (2.16) that

$$\big|(\psi - v)(X)\big| < C_1 \epsilon g(X) \qquad (X \in K \cup E_1) \tag{2.17}$$

and

$$\big|\psi(X)\big| < C_1 \epsilon g(X) \qquad (X \in K \cup E_2). \tag{2.18}$$

The remainder of the proof is concerned with approximating ψ on $K \cup E_1 \cup E_2$ by a function u in $\mathcal{I}(\Omega)$.

Let S denote the (finite) set of singularities of v in ω_1, and define $S_\delta = \{X : \operatorname{dist}(X, S) \le \delta\}$. By choosing δ to be sufficiently small, we can arrange that S_δ is the disjoint union of closed balls contained in ω_1. Let $W = \omega_1 \cup V \cup (\Omega \backslash \overline{\omega}_2)$. Using Lemma 2.4 (with $(\omega_1 \cup V) \backslash S_\delta$ in place of U) and the fact that $\psi = 0$ on $\mathbf{R}^n \backslash \omega_2$, we obtain

$$\psi(X) = s_1(X) + s_2(X) + s_3(X) \qquad (X \in W \backslash S_\delta), \tag{2.19}$$

where

$$s_1(X) = -c_n \int_{W \backslash S_\delta} G(X,Y) \Delta\psi(Y) d\lambda(Y),$$

$$s_2(X) = -c_n \int_{\partial W} \left\{ \psi(Y) \frac{\partial}{\partial n_Y} G(X,Y) - G(X,Y) \frac{\partial \psi}{\partial n_Y}(Y) \right\} d\sigma(Y)$$

and

$$s_3(X) = c_n \int_{\partial S_\delta} \left\{ \psi(Y) \frac{\partial}{\partial n_Y} G(X,Y) - G(X,Y) \frac{\partial \psi}{\partial n_Y}(Y) \right\} d\sigma(Y). \quad (2.20)$$

Since $\Delta\psi = \Delta v = 0$ on $\omega_1 \backslash S$, we see that s_1 is independent of δ, and the same must therefore also be true of s_3. Thus (2.20) defines a harmonic function on $\Omega \backslash S$. Let

$$s_4(X) = -2c_n \int_{\partial W} v(Y) G(X,Y) \frac{\partial \phi}{\partial n_Y}(Y) d\sigma(Y) \qquad (X \in \Omega).$$

Since $|\nabla\phi| = 0$ on ∂S_δ, we can apply Green's first identity to obtain

$$\begin{aligned} s_4(X) = -2c_n \int_{W \backslash S_\delta} \{ & v(Y) G(X,Y) \Delta\phi(Y) \\ & + \langle \nabla_2 \{ v(Y) G(X,Y) \}, \nabla\phi(Y) \rangle \} \, d\lambda(Y), \\ = -2c_n \int_{W \backslash S_\delta} \{ & v(Y) G(X,Y) \Delta\phi(Y) + G(X,Y) \langle \nabla v(Y), \nabla\phi(Y) \rangle \\ & + v(Y) \langle \nabla_2 G(X,Y), \nabla\phi(Y) \rangle \} \, d\lambda(Y) \end{aligned} \quad (2.21)$$

whenever $X \in W \backslash S_\delta$, where $\langle \cdot, \cdot \rangle$ denotes the usual inner product on $\mathbf{R}^n$. Noting that $\Delta\psi = v\Delta\phi + 2\langle \nabla v, \nabla\phi \rangle$ on $W \backslash S$, we obtain

$$\begin{aligned} s_1(X) &= s_4(X) \\ &+ c_n \int_{W \backslash S_\delta} v(Y) \{ G(X,Y) \Delta\phi(Y) + 2 \langle \nabla_2 G(X,Y), \nabla\phi(Y) \rangle \} d\lambda(Y). \end{aligned}$$

Hence, letting $\delta \to 0+$ and noting that $\Delta\phi = 0 = |\nabla\phi|$ on $\omega_1 \cup (\mathbf{R}^n \backslash \omega_2)$, we see from (2.19) that

$$\begin{aligned} \psi(X) &= s_2(X) + s_3(X) + s_4(X) \\ &+ c_n \int_V v(Y) \{ G(X,Y) \Delta\phi(Y) + 2 \langle \nabla_2 G(X,Y), \nabla\phi(Y) \rangle \} \, d\lambda(Y) \\ & \qquad\qquad\qquad\qquad\qquad\qquad\qquad\qquad\qquad (X \in W \backslash S). \end{aligned}$$

It follows from (2.16) that

$$\left|(s_2 + s_3 + s_4)(X) - \psi(X)\right| \le C_2\epsilon \int_V \left\{G(X,Y) + \left|\nabla_2 G(X,Y)\right|\right\} d\lambda(Y)$$
$$\le C_2\epsilon F(X) \qquad (X \in W\backslash S), \tag{2.22}$$

where C_2 is a constant independent of v (and hence of u_1) and

$$F(X) = \int_U \left\{G(X,Y) + \left|\nabla_2 G(X,Y)\right|\right\} d\lambda(Y).$$

Letting ω denote an Ω-bounded open set such that $\overline{U} \subset \omega$, we see from Lemma 2.2 that

$$F(X) \le C_3 g(X) \qquad (X \in \Omega\backslash\omega). \tag{2.23}$$

Also, since F is continuous on Ω,

$$F(X) \le C_4 g(X) \qquad (X \in \overline{\omega}). \tag{2.24}$$

Combining (2.22)–(2.24) we obtain

$$\left|(s_2 + s_3 + s_4)(X) - \psi(X)\right| \le C_5\epsilon g(X) \qquad (X \in W\backslash S), \tag{2.25}$$

where C_5 is independent of u_1. We noted above that $s_3 \in \mathcal{H}(\Omega\backslash S)$. Also, if ω' is an Ω-bounded open set for which

$$\partial(\omega_1 \cup V) \cap \operatorname{supp}\psi \subset \omega' \text{ and } \overline{\omega'} \cap (K \cup E_1 \cup E_2) = \emptyset,$$

then there exist w_2 and w_4 in $\mathcal{I}(\Omega)$ such that

$$\left|(w_k - s_k)(X)\right| < \epsilon g(X) \qquad (X \in \Omega\backslash\omega'; k = 2, 4): \tag{2.26}$$

when $k = 2$ this follows from Lemma 2.3, and similar reasoning deals with the case $k = 4$. If we now define $u = w_2 + s_3 + w_4$, then $u \in \mathcal{I}(\Omega)$, and we see from (2.25) and (2.26) that

$$\left|(u - \psi)(X)\right| < (C_5 + 2)\epsilon g(X) \qquad (X \in W\backslash(\omega' \cup S)).$$

Combining this with (2.15), (2.17) and (2.18) we obtain

$$\left|(u - u_1)(X)\right| < C_6\epsilon g(X) \qquad (X \in K \cup E_1)$$

(any points in $S \cap E_1$ must be removable singularities of $u - u_1$) and

$$\left|u(X)\right| < C_6\epsilon g(X) \qquad (X \in K \cup E_2),$$

where $C_6 = C_1 + C_5 + 3$. The theorem is now proved.

Notes

The above exposition is based on that given in Armitage and Goldstein [AG3], which in turn made use of ideas from Gauthier, Goldstein and Ow [GGO1], [GGO2] and from the lecture notes by Gauthier and Hengartner [GH].

3 Approximation on Relatively Closed Sets

3.1 Introduction

The following is Arakelyan's generalization of Mergelyan's Theorem (see §1.1) to non-compact sets. It can be found in [Ara1] or [Ara2].

Arakelyan's Theorem (1968). *Let Ω be an open set in* $\mathbf{C}$ *and E be a relatively closed subset of Ω. The following are equivalent:*
(a) for each f in $C(E) \cap \mathrm{Hol}(E^\circ)$ and each positive number ϵ, there exists g in $\mathrm{Hol}(\Omega)$ such that $|g - f| < \epsilon$ on E;
(b) $\Omega^* \backslash E$ is connected and locally connected.

The above local connectedness condition will be discussed in §3.2. Its first appearance (at least, in an equivalent form) in the context of holomorphic approximation occurs in early work of Alice Roth which is not as well known as it should be. It is remarkable that, as early as 1938, Roth [Rot1] had shown that (when $\Omega = \mathbf{C}$) condition (b) above is sufficient for uniform approximation of functions in $\mathrm{Hol}(E)$ by entire holomorphic functions. (See [Rot2] for the generalization to other choices of Ω.) Of course, Arakelyan's Theorem is an improvement of Roth's result.

This chapter presents corresponding results for uniform approximation by harmonic functions on relatively closed sets. In fact, we will obtain generalizations of Theorems 1.3, 1.7, 1.10, 1.15, and Corollary 1.16. Further, it will be shown that, whenever uniform approximation is possible, something rather better is also true (at least, in most cases). The main results are Theorems 3.15, 3.17 and 3.19.

Throughout this chapter Ω will denote a connected open set in $\mathbf{R}^n$ and E will denote a relatively closed proper subset of Ω. In §§3.3–3.4 we will impose the additional hypothesis that Ω is Greenian.

3.2 Local Connectedness

Arakelyan's Theorem (see §3.1) shows that (when $n = 2$) local connectedness of $\Omega^*\backslash E$ plays a vital role in the theory of holomorphic approximation. This notion is also useful in the preliminary results of this chapter, so we briefly discuss it here.

Recall that a topological space is called *locally connected* if, for each point X in the space and each neighbourhood ω of X, there is a connected neighbourhood ω' of X such that $\omega' \subseteq \omega$. The set $\Omega^*\backslash E$ can only fail to satisfy this condition in the case where $X = \mathcal{A}$. Thus $\Omega^*\backslash E$ is locally connected if and only if for each compact subset K of Ω there is a compact subset L of Ω which contains every Ω-bounded component of $\Omega\backslash(E \cup K)$. We give below an example of a pair (Ω, E) such that $\Omega^*\backslash E$ is connected but not locally connected.

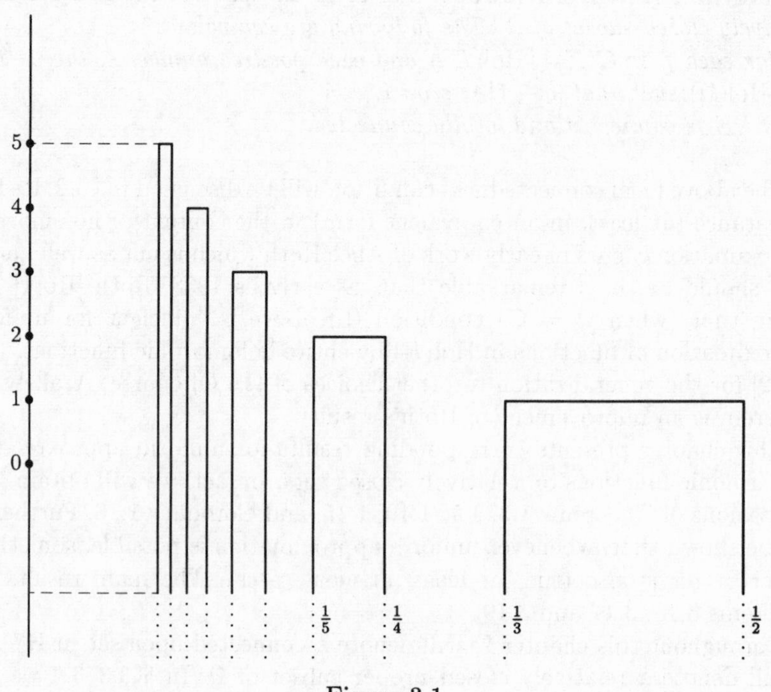

Figure 3.1

Example 3.1. If

$$S = \bigcup_{k=1}^{\infty} \left\{ (x_1, x_2) : \frac{1}{2k+1} < x_1 < \frac{1}{2k} \text{ and } x_2 < k \right\}$$

(see Figure 3.1), then $(\mathbf{R}^2)^* \backslash \partial S$ is connected. However, $(\mathbf{R}^2)^* \backslash \partial S$ is not locally connected: the set $(\mathbf{R}^2)^* \backslash (\partial S \cup K)$, where $K = [0,1] \times \{0\}$, is a neighbourhood of $\mathcal{A}$ in $(\mathbf{R}^2)^* \backslash \partial S$ which does not contain any connected neighbourhood of $\mathcal{A}$. (There is no compact set L which contains all the bounded components of $(\mathbf{R}^2)^* \backslash (\partial S \cup K)$.) The above set S is sometimes referred to as an *Arakelyan glove.*

Let $Y \in \Omega$. If $f : [0, +\infty) \to \Omega$ is a continuous function such that $f(0) = Y$ and $f(t) \to \mathcal{A}$ as $t \to +\infty$, then we call the image $f([0, +\infty))$ a *path connecting* Y *to* $\mathcal{A}$. By an *exhaustion of* Ω we mean a sequence (K_m) of compact sets such that $\cup_m K_m = \Omega$, and such that $K_m \subset K_{m+1}^\circ$ for each m. The following simple lemma will be useful.

Lemma 3.2. *Suppose that* $\Omega^* \backslash E$ *is connected and locally connected. Then there is an exhaustion* (K_m) *of* Ω *such that both* $\Omega^* \backslash K_m$ *and* $\Omega^* \backslash (E \cup K_m)$ *are connected for each* m.

Proof. Let $Q \in \Omega$, let $L_1 = \{Q\}$, and let (L_m) be an exhaustion of Ω. We inductively define a new exhaustion (K_m) as follows. Let $K_1 = L_1$. Given K_m, choose l large enough to ensure that $K_m \subset L_l^\circ$, let $\{V_k\}$ denote the collection of Ω-bounded components of $\Omega \backslash (E \cup L_l)$ and define $K_{m+1} = \widehat{F}$, where $F = \overline{(\cup_k V_k)} \cup L_l$. The local connectedness of $\Omega^* \backslash E$ shows that F, and thus also K_{m+1}, are compact subsets of Ω. Clearly $K_m \subset L_l^\circ \subset K_{m+1}^\circ$, and neither $\Omega \backslash K_{m+1}$ nor $\Omega \backslash (E \cup K_{m+1})$ have Ω-bounded components. Finally, the sequence (K_m) constructed in this way has the property that $\cup_m K_m = \Omega$, since $\cup_m L_m = \Omega$.

We observe that, if (K_m) is as in Lemma 3.2 and $Y \in \Omega \backslash (E \cup K_m)$, then there is a path connecting Y to $\mathcal{A}$ which is contained in $\Omega \backslash (E \cup K_m)$. This must be the case because the component of $\Omega \backslash (E \cup K_m)$ to which Y belongs is not Ω-bounded, and so there is a continuous function $f : [0,1] \to \Omega \backslash (E \cup K_m)$ such that $f(0) = Y$ and $f(1) \in \Omega \backslash (E \cup K_{m+1})$. An inductive argument now establishes the claim.

3.3 Pole Pushing

We saw in Theorem 1.7 that, if K is a compact subset of an open set Ω such that $\Omega^*\backslash K$ is connected, then any function in $\mathcal{H}(K)$ can be uniformly approximated (on K) by functions in $\mathcal{H}(\Omega)$. This was done by first approximating a given member of $\mathcal{H}(K)$ by a function h in $\mathcal{I}(\Omega)\cap\mathcal{H}(K)$, and then using a pole pushing argument to approximate h by a member of $\mathcal{H}(\Omega)$. It is the latter step which we generalize below to deal with approximation on unbounded sets.

In this section we assume that Ω is a connected open set in $\mathbf{R}^n$ with Green function $G(\,\cdot\,,\,\cdot\,)$, and define $g(\,\cdot\,) = \min\{1, G(Q,\,\cdot\,)\}$, where Q is some fixed point of Ω. As usual, E denotes a relatively closed proper subset of Ω.

Theorem 3.3. *If $\Omega^*\backslash E$ is connected and locally connected, then for each u in $\mathcal{I}(\Omega)\cap\mathcal{H}(E)$ and each positive number ϵ, there exists v in $\mathcal{H}(\Omega)$ such that*

$$\left|(v-u)(X)\right| < \epsilon g(X) \qquad (X \in E).$$

Before proving Theorem 3.3 we require a definition and a lemma. If $u \in \mathcal{I}(\Omega)$ and Y is a singularity of u, then an open subset T of Ω is called a *tract for u and Y* if $u \in \mathcal{H}(T\backslash\{Y\})$ and T contains a path connecting Y to $\mathcal{A}$.

Lemma 3.4. *Suppose that $u \in \mathcal{I}(\Omega)$ and $\epsilon > 0$, and that Y_0 is a singularity of u. If T is a tract for u and Y_0, then there exists v in $\mathcal{I}(\Omega)\cap\mathcal{H}(T)$ such that*

$$\left|(v-u)(X)\right| < \epsilon g(X) \qquad (X \in \Omega\backslash T).$$

Proof of Lemma. Using the Laurent expansion for u about Y_0, we can write $u = u_0 + v_0$, where $u_0 \in \mathcal{H}(\mathbf{R}^n\backslash\{Y_0\})$ and $v_0 \in \mathcal{H}(\{Y_0\})$. Thus $v_0 \in \mathcal{I}(\Omega)\cap\mathcal{H}(T)$. We choose a sequence $\big(B(Y_k,r_k)\big)_{k\geq 0}$ of balls in T such that $Y_{k-1} \in B(Y_k,r_k)$ and $r_k \to 0$, and such that $Y_k \to \mathcal{A}$. By repeated application of Lemma 2.6 we obtain a sequence $(u_k)_{k\geq 0}$ of functions such that $u_k \in \mathcal{H}(\Omega\backslash\{Y_k\})$ and

$$\left|(u_k-u_{k-1})(X)\right| < 2^{-k}\epsilon g(X) \qquad (X \in \Omega\backslash B(Y_k,r_k))$$

for each k in $\mathbf{N}$. Clearly (u_k) converges locally uniformly on Ω. We can thus define $v = v_0 + \lim_{k\to\infty} u_k$, and observe that $v \in \mathcal{I}(\Omega)\cap\mathcal{H}(T)$ and

$$\left|(v-u)(X)\right| \leq \sum_{k=1}^{\infty}\left|(u_k-u_{k-1})(X)\right| < \epsilon g(X) \qquad (X \in \Omega\backslash T),$$

as required.

Proof of Theorem 3.3. Let $\epsilon > 0$, let $u \in \mathcal{I}(\Omega) \cap \mathcal{H}(E)$, and let $Y_1, Y_2, \ldots$ denote the singularities of u. Thus $Y_k \in \Omega\backslash E$ for each k. It is clear from the connectedness hypotheses and the observations following Lemma 3.2 that, for each k, we can find a path connecting Y_k to $\mathcal{A}$ which lies in $\Omega\backslash E$, and that this can be done in such a way that each compact subset of Ω intersects only finitely many of these paths. Hence, for each k, we can find a tract T_k for u and Y_k such that $T_k \subset \Omega\backslash E$, and can ensure that each compact subset of Ω intersects only finitely many of these tracts.

Let $u_0 = u$. By repeated application of Lemma 3.4 we obtain a sequence $(u_k)_{k\geq 0}$ of functions such that the singularities of (u_k) are $\{Y_j : j \geq k+1\}$, and

$$\left|(u_k-u_{k-1})(X)\right| < 2^{-k}\epsilon g(X) \qquad (X \in \Omega\backslash T_k; k \in \mathbf{N}).$$

It follows from the local finiteness of the family $\{T_k : k \geq 1\}$ that (u_k) converges locally uniformly on Ω to a function v in $\mathcal{H}(\Omega)$. Further, since $T_k \cap E = \emptyset$ for each k, we see that

$$\left|(v-u)(X)\right| \leq \sum_{k=1}^{\infty}\left|(u_k-u_{k-1})(X)\right| < \epsilon g(X) \qquad (X \in E).$$

3.4 A Sufficient Condition for Runge Approximation

Let Ω, E and g be as in §3.3. In this section we improve Theorem 3.3 by showing that u need only belong to $\mathcal{H}(E)$. This will be done by means of the fusion result in Chapter 2.

Theorem 3.5. *For each u in $\mathcal{H}(E)$ and each positive number ϵ, there exists v in $\mathcal{I}(\Omega) \cap \mathcal{H}(E)$ such that*

$$\left|(v-u)(X)\right| < \epsilon g(X) \qquad (X \in E). \tag{3.1}$$

Corollary 3.6 *If $\Omega^*\backslash E$ is connected and locally connected, then for each u in $\mathcal{H}(E)$ and each positive number ϵ, there exists v in $\mathcal{H}(\Omega)$ such that (3.1) holds.*

The Corollary follows immediately from Theorems 3.5 and 3.3. Before proving Theorem 3.5 we present an example to show that the error bound $\epsilon g(X)$ in (3.1) cannot, in general, be improved.

Example 3.7. Let $\Omega = \mathbf{R}^{n-1} \times (0, +\infty)$, let $E = \overline{B(O,1)} \cap \Omega$, and let

$$u(X) = \sum_{k=1}^{\infty} 2^{-k} \phi_n\Big(\big|X - (0, \dots, 0, 2 + 2^{-k})\big|\Big) \qquad (X \in \Omega).$$

Then $u \in \mathcal{H}(E)$. Now suppose that there exists v in $\mathcal{I}(\Omega) \cap \mathcal{H}(E)$ such that

$$\big|(v-u)(X)\big| = o(g(X)) \qquad (X \to Y; X \in E)$$

for all Y in $\overline{B(O,1)} \cap \partial\Omega$. The Green function for Ω is given by

$$G(X,Y) = \phi_n(|X-Y|) - \phi_n\Big(\big|X - (y_1, \dots, y_{n-1}, -y_n)\big|\Big),$$

where $Y = (y_1, y_2, \dots, y_n)$. Simple estimates now show that

$$\big|(v-u)(X)\big| = o(x_n) \qquad (X \to Y; X \in E)$$

for all Y in $B(O,1) \cap \partial\Omega$. Thus $v - u$ and $\partial(v-u)/\partial x_n$ vanish on the latter set. It follows that $v = u$ on E and hence on Ω. This leads to a contradiction, in view of the infinite number of singularities of u in the compact set $\{0\}^{n-1} \times [2,3]$.

Proof of Theorem 3.5. Let $u \in \mathcal{H}(E)$, let $\epsilon > 0$, and let (K_m) be an exhaustion of Ω. Also, let $F_m = K_m \cap E$, and let C_m be the constant of Theorem 2.7 corresponding to the assignments $K = F_{m+1}, E_1 = K_m$ and $E_2 = E \backslash K_{m+1}^{\circ}$. We may assume that $C_m \geq 1$, and we define (δ_m) to be a decreasing sequence of positive numbers such that $\delta_m \leq 2^{-m-2}\epsilon / C_m$. It follows from Lemma 1.8 that, for each m, there exists q_m in $\mathcal{I}(\mathbf{R}^n)$ such that

$$\big|(q_m - u)(X)\big| < \delta_m g(X) \qquad (X \in F_{m+1}), \tag{3.2}$$

whence $q_m, q_{m+1} \in \mathcal{H}(F_{m+1})$ and

$$\big|(q_{m+1} - q_m)(X)\big| < 2\delta_m \qquad (X \in F_{m+1}).$$

Thus we can apply Theorem 2.7 to obtain r_m in $\mathcal{I}(\Omega)$ such that

$$\big|(r_m - q_m)(X)\big| \leq 2\delta_m C_m g(X) \leq 2^{-m-1}\epsilon g(X) \qquad (X \in K_m \cup F_{m+1}) \tag{3.3}$$

and

$$\big|(r_m - q_{m+1})(X)\big| \leq 2^{-m-1}\epsilon g(X) \qquad (X \in E). \tag{3.4}$$

It follows from (3.3) that $r_m - q_m$, suitable defined at its removable singularities, belongs to $\mathcal{H}(K_m^{\circ})$, and that the series

$$\sum_{k=m}^{\infty} (r_k - q_k)(X)$$

defines a harmonic function on K_m°. Thus, if we define

$$v = q_1 + \sum_{k=1}^{\infty} (r_k - q_k),$$

we see that $v \in \mathcal{I}(K_m^\circ)$ for any m, and hence $v \in \mathcal{I}(\Omega)$.

If $X \in F_m$, then

$$\begin{aligned}
|(v-u)(X)| &\leq \sum_{k=1}^{m-1} |(r_k - q_{k+1})(X)| + |(q_m - u)(X)| + \sum_{k=m}^{\infty} |(r_k - q_k)(X)| \\
&< \sum_{k=1}^{\infty} 2^{-k-1} \epsilon g(X) + 2^{-m-2} \epsilon g(X) \\
&< \epsilon g(X),
\end{aligned}$$

using (3.2)–(3.4). Since this holds for each m, we have $|v - u| < \epsilon g$ on E, and thus $v \in \mathcal{H}(E)$. The theorem is now established.

3.5 Relating the Error to the Set E

From now on Ω denotes a connected open set in $\mathbf{R}^n$, and E is a relatively closed proper subset of Ω. The main result of this section is Theorem 3.11 below. However, we begin by recording three easy consequences of Corollary 3.6.

Corollary 3.8. *If $\Omega^* \backslash E$ is connected and locally connected, then for each u in $\mathcal{H}(E)$ and each positive number ϵ, there exists v in $\mathcal{H}(\Omega)$ such that $|v - u| < \epsilon$ on E.*

Proof. If Ω is Greenian, then the conclusion of the Corollary is immediate from Corollary 3.6, since $g \leq 1$. It remains to deal with the case where $n = 2$ and Ω is not Greenian. Let $Y_0 \in \Omega \backslash E$ and choose a positive number r_0 such that $\overline{B(Y_0, r_0)} \subset \Omega \backslash E$. If we define $\Omega_0 = \Omega \backslash \overline{B(Y_0, r_0)}$, then Ω_0 possesses a Green function, so there exists w in $\mathcal{H}(\Omega_0)$ such that $|w - u| < \epsilon/2$. It follows from the connectedness hypotheses that there is a path connecting Y_0 to $\mathcal{A}$. Hence there is a sequence $\left(\overline{B(Y_k, r_k)}\right)$ of balls in $\Omega \backslash E$ such that $r_k \to 0$ and $Y_{k-1} \in B(Y_k, r_k)$ for each k, and such that $Y_k \to \mathcal{A}$. Repeated application of Lemma 1.9 (with ϵ replaced by $2^{-k-1}\epsilon$) yields (in the limit) a harmonic function v in Ω such that $|v - w| < \epsilon/2$ on E, whence $|v - u| < \epsilon$ on E.

A useful application of Corollary 3.8 is the following analogue of the Mittag-Leffler theorem. It can also be proved directly from Theorem 1.7 (cf. [Con, p. 205]).

Corollary 3.9. *Let (X_m) be a sequence of distinct points in Ω such that $X_m \to \mathcal{A}$, and for each m let u_m be harmonic on a deleted neighbourhood of X_m. Then there is a harmonic function v on $\Omega\backslash\{X_m : m \in \mathbf{N}\}$ such that $v - u_m$ has a removable singularity at X_m, for each m.*

Proof. For each m let B_m be a closed ball centred at X_m such that $u_m \in \mathcal{H}(B_m\backslash\{X_m\})$. Further, we arrange that these balls are pairwise disjoint. Let

$$\Omega_0 = \Omega\backslash\{X_m : m \in \mathbf{N}\}, \qquad E = \bigcup_{m=1}^{\infty} (B_m\backslash\{X_m\})$$

and let $u = u_m$ on a neighbourhood of $B_m\backslash\{X_m\}$. Then E is a relatively closed subset of Ω_0, and $(\Omega_0)^*\backslash E$ is connected and locally connected. It follows from Corollary 3.8 that there exists v in $\mathcal{H}(\Omega_0)$ such that $|v-u| < 1$ on E. Thus $v - u_m$ must have a removable singularity at X_m, for each m.

If $A \subseteq \mathbf{R}^n$, then $\mathcal{S}(A)$ will denote the collection of all functions which are superharmonic on some open set containing A. Further $\mathcal{S}^+(A)$ (resp. $\mathcal{H}^+(A)$) will denote the collection of positive functions in $\mathcal{S}(A)$ (resp. $\mathcal{H}(A)$). Thus $u \in \mathcal{S}^+(A)$ if and only if u is defined, positive and superharmonic on some open set which contains A.

Corollary 3.10. *If $\Omega^*\backslash E$ is connected and locally connected, then for each u in $\mathcal{H}(E)$ and each s in $\mathcal{S}^+(\Omega)$, there exists v in $\mathcal{H}(\Omega)$ such that $|v-u| < s$ on E.*

Proof. Let $s \in \mathcal{S}^+(\Omega)$. If Ω is not Greenian, then $n = 2$ and s is constant by Myrberg's theorem (see [Helm, Theorem 8.33]). The conclusion then follows from Corollary 3.8. If Ω possesses a Green function $G(\,\cdot\,,\,\cdot\,)$, then let B be a closed ball in Ω with centre Q, choose a number a such that

$$a > \sup\Big(\{G(Q,X) : X \in \Omega\backslash B\} \cup \{1\}\Big),$$

and define

$$b = \inf\left\{\frac{s(X)}{\min\{a, G(Q,X)\}} : X \in B\right\}.$$

It follows from the minimum principle and the harmonicity of the function $\min\{a, G(Q,\,\cdot\,)\}$ on $\Omega\backslash B$ that

$$s(X) \geq b \min\{a, G(Q,X)\} \geq b \min\{1, G(Q,X)\} \qquad (X \in \Omega).$$

Hence the conclusion follows in this case from Corollary 3.6.

In fact, Corollary 3.10 can be significantly improved to yield the following result, which will be proved in §3.6.

Theorem 3.11 *If $\Omega^* \backslash E$ is connected and locally connected, then for each u in $\mathcal{H}(E)$ and each s in $\mathcal{S}^+(E)$, there exists v in $\mathcal{H}(\Omega)$ such that $0 < v - u < s$ on E.*

The advantage gained in Theorem 3.11 is that we obtain a better approximation if restrictions are imposed on the set E. A simple example is given below to illustrate this comment and also the sharpness of the speed of approximation.

Example 3.12. Let $\Omega = \mathbf{R}^n$, let $\omega = (-1,1)^{n-1} \times \mathbf{R}$ and $\alpha = (\pi/2)(n-1)^{1/2}$. Then:
(i) given any closed set E such that $E \subset \omega$ and the hypotheses of Theorem 3.11 are satisfied, any u in $\mathcal{H}(E)$ and any positive number ϵ, there exists v in $\mathcal{H}(\mathbf{R}^n)$ such that

$$0 \leq v(X) - u(X) < \epsilon \exp(-\alpha |x_n|) \qquad (X = (x_1, \ldots, x_n) \in E);$$

and
(ii) the above statement becomes false if α is replaced by any larger number.

Details of Example. To show that (i) holds, we define $\omega_\delta = (-\delta, \delta)^{n-1} \times \mathbf{R}$ and

$$s_\delta(X) = \cos(\pi x_1/(2\delta)) \cdots \cos(\pi x_{n-1}/(2\delta)) \exp(-\alpha |x_n|/\delta).$$

Then $s_\delta \in \mathcal{S}^+(\omega_\delta)$, since s_δ is the minimum of two positive harmonic functions on ω_δ. (In fact, s_δ is a potential on ω_δ.) Assertion (i) above now follows immediately from Theorem 3.11, since $s_1 \in \mathcal{S}^+(E)$.

To prove (ii), let $\beta > \alpha$ and choose δ in $(0,1)$ close enough to 1 to ensure that $\alpha/\delta < \beta$. Let $P = (1, 0, \ldots, 0)$ and $u(X) = \phi_n(|X - P|)$, so that $u \in \mathcal{H}(E)$, where $E = \overline{\omega_{(1+\delta)/2}}$. Now suppose that $v \in \mathcal{H}(\mathbf{R}^n)$ and $v - u \geq 0$ on E. The minimum principle ensures that $v - u > 0$ on $\omega_{(1+\delta)/2}$, since $(v-u)(P) = +\infty$. If we define

$$c = \inf \left\{ \frac{v(X) - u(X)}{s_\delta(X)} : X = (x_1, \ldots, x_{n-1}, 0) \in \omega_\delta \right\},$$

then $v - u \geq cs_\delta$ on ω_δ by the Maria-Frostman domination principle (see [Helm, Theorem 8.43]). Thus

$$v(0,\ldots,0,x_n) - u(0,\ldots,0,x_n) \geq c\exp(-\alpha|x_n|/\delta) \qquad (x_n \in \mathbf{R}),$$

and (ii) follows, since $\alpha/\delta < \beta$.

3.6 Proof of Theorem 3.11

Let $u \in \mathcal{H}(E)$ and $s \in \mathcal{S}^+(E)$. Thus there are open sets ω_1, ω_2, contained in Ω, which contain E such that u is harmonic on ω_1 and s is positive and superharmonic on ω_2. It is enough to deal with the case where s is actually harmonic on ω_2. To see this we observe that if, on each component W of ω_2, we replace s by its reduced function relative to a closed ball contained in $W\backslash E$, then the resultant function belongs to $\mathcal{H}^+(E)$ and does exceed s. Next, it is enough to show that there exists v in $\mathcal{H}(\Omega)$ such that $|v-u| < s$ on E. For, if this were done, we could then obtain v in $\mathcal{H}(\Omega)$ such that $|v-(u+s/2)| < s/2$, and hence $0 < v-u < s$ on E. Also, we may assume that $\inf_E s = 0$, for otherwise we may appeal to Corollary 3.8.

Let ω be an open set such that $E \subset \omega$ and $\overline{\omega} \cap \Omega \subset \omega_1 \cap \omega_2$, let (K_m) be an exhaustion of Ω as in Lemma 3.2, and let a denote the infimum of the values of s on the set $\overline{\omega} \cap K_1$. (We can arrange that this latter set is non-empty.) For each k in $\mathbf{N}$ we define the sets

$$D_k = \{X \in \omega_2 : s(X) \leq 2^{-k}a\} \text{ and } \Omega_k = \Omega\backslash(\partial\omega \cap D_k),$$

and let $m(k)$ be the largest integer m such that K_m is disjoint from $\overline{\omega} \cap D_k$. (The existence of $m(k)$ follows from the assumption that $\inf_E s = 0$.) We observe that, for a fixed choice of m, the set $K_m \cap \overline{\omega} \cap D_k$ is empty for all sufficiently large values of k. Thus $m(k) \to \infty$ as $k \to \infty$.

We know from Lemma 3.2 that $\Omega^*\backslash(E \cup K_{m(k)})$ is connected, and the same is therefore true of $(\Omega_{k+1})^*\backslash(E \cup K_{m(k)})$. The latter set must also be locally connected. For, if this were not the case, then there would be a compact subset L of Ω_{k+1} (and hence of Ω) for which the Ω_{k+1}-bounded components $\{V_j\}$ of $\Omega_{k+1}\backslash(E \cup K_{m(k)} \cup L)$ do not have an Ω_{k+1}-bounded union. Thus there is a sequence of points (X_l) such that each X_l belongs to some V_j, and such that (X_l) converges to the Alexandroff point for Ω_{k+1}. However, it follows from the fact that $\Omega^*\backslash E$ is locally connected that $\cup_j V_j$ is Ω-bounded, and so there is a subsequence of (X_l) which converges to some point of E. Since $E \cap \partial\Omega_{k+1} \subseteq E \cap \partial\omega = \emptyset$, this leads to the desired contradiction.

We now define a positive superharmonic function on Ω_k by

$$s_k(X) = \begin{cases} s(X) & (X \in \omega \cap D_k) \\ 2^{-k}a & (\text{elsewhere in } \Omega_k). \end{cases}$$

We observe that $s_k = \min\{s, 2^{-k}a\}$ on ω, and that $s_{k+1} \le s_k$ on Ω_k. Next, we inductively define a sequence (v_k) of harmonic functions as follows. By Corollary 3.10 there exists v_1 in $\mathcal{H}(\Omega_1)$ such that $|v_1 - u| < 2^{-1}s_1$ on E. Given v_k in $\mathcal{H}(\Omega_k)$ we use Corollary 3.10 and the observations of the above paragraph to obtain v_{k+1} in $\mathcal{H}(\Omega_{k+1})$ such that

$$|v_{k+1} - v_k| < 2^{-k-1}s_{k+1} \quad \text{on } E \cup K_{m(k)}.$$

Thus we obtain a harmonic function on Ω by defining $v = \lim_{k\to\infty} v_k$. Further, on E we have

$$\begin{aligned} |v_k - u| &\le \sum_{j=2}^{k} |v_j - v_{j-1}| + |v_1 - u| \\ &< \sum_{j=2}^{k} 2^{-j}s_j + 2^{-1}s_1 < s, \end{aligned}$$

and it follows that $|v - u| < s$ on E, as required.

3.7 A Necessary Condition for Runge Approximation

We will say that the pair (Ω, E) *satisfies the* (K, L)*-condition* if, for each compact subset K of Ω, there is a compact subset L of Ω which contains every Ω-bounded component of $\Omega\backslash(E \cup K)$ whose closure intersects K.

Example 3.13. If

$$S = \bigcup_{k=1}^{\infty} \left\{(x_1, x_2) : \frac{1}{2k+1} < x_1 < \frac{1}{2k} \text{ and } 0 < x_2 < k\right\},$$

then $(\mathbf{R}^2, \partial S)$ does not satisfy the (K, L)-condition: if we let $K = [0,1]\times\{0\}$, then for each k the set V_k given by $V_k = \left((2k+1)^{-1}, (2k)^{-1}\right) \times (0, k)$ is an Ω-bounded component of $\mathbf{R}^2\backslash(\partial S \cup K)$ whose closure intersects K.

It is not hard to see that, if (Ω, E) satisfies the (K, L)-condition, then $\Omega^*\backslash\widehat{E}$ is locally connected. However, $\Omega^*\backslash E$ need not be locally connected: for, if

$$E = \bigcup_{k=1}^{\infty} \partial B\left((3k, 0, \ldots, 0), 1\right),$$

then $(\mathbf{R}^n, E)$ satisfies the (K, L)-condition, but $(\mathbf{R}^n)^* \backslash E$ is not locally connected.

The next result shows that the (K, L)-condition is necessary if we wish to be able to approximate functions in $\mathcal{H}(E)$ by functions in $\mathcal{H}(\Omega)$. It is stated in a form which will be useful also in later chapters.

Theorem 3.14. *Suppose that, for each h in $\mathcal{H}(E)$, there exists v in $\mathcal{S}(\Omega)$ such that $v \geq h$ on E. Then (Ω, E) satisfies the (K, L)-condition.*

Proof. To prove this, suppose that the (K, L)-condition fails to hold. Then there is a compact subset K of Ω, a sequence (V_k) of distinct Ω-bounded components of $\Omega \backslash (E \cup K)$, and two sequences (X_k), (Y_k) of points, such that $X_k, Y_k \in V_k$ for each k, and such that $X_k \to \mathcal{A}$ and (Y_k) converges to some point Y_0 in K. Now let U be an Ω-bounded open set which contains K and let U_0 be the component of U which contains Y_0. By deleting the first few members of the sequence (V_k) we can arrange that $V_k \cap U_0 \neq \emptyset$ for each k. We define

$$a_k = \mu_{V_k, X_k}(U_0 \cap \partial V_k) \qquad (k \in \mathbf{N}), \tag{3.5}$$

where $\mu_{V,X}$ denotes harmonic measure for an open set V and a point X in V. If $a_k = 0$, then (see [Doo, 1.VIII.5(b)]) there is a superharmonic function u_1 on V_k with limit $+\infty$ at each point of $U_0 \cap \partial V_k$. Hence the function

$$u_2(X) = \begin{cases} u_1(X) & (X \in U_0 \cap V_k) \\ +\infty & (X \in U_0 \backslash V_k) \end{cases}$$

is lower semicontinuous and super-meanvalued on U_0. This is impossible, since $U_0 \cap V_{k+1}$ is a non-polar subset of $U_0 \backslash V_k$. Thus $a_k > 0$ for each k.

It follows from Corollary 3.9 that there is a harmonic function h on the set Ω_1, where $\Omega_1 = \Omega \backslash \{X_k : k \in \mathbf{N}\}$, such that, for each k, the function

$$h(X) + a_k^{-1} \phi_n(|X - X_k|)$$

has a harmonic extension to $\Omega_1 \cup \{X_k\}$. By hypothesis there exists a superharmonic function v on Ω such that $v \geq h$ on E. Since $v - h$ is superharmonic on Ω, we can define b to be a negative lower bound for $v - h$ on $\overline{U}$, and then define W to be the open subset of Ω where $v - h > b - 1$. It follows from the minimum principle, and the fact that $K \subset U$, that $v - h \geq b$ on each $\overline{V_k}$, and so $\cup_k \overline{V_k} \subseteq W$. Also, $\overline{U} \subset W$. Clearly the function u defined by $u(X) = v(X) - h(X) - b + 1$ is positive and superharmonic on W, and satisfies $\nu_u(\{X_k\}) \geq a_k^{-1}$ for each k, where ν_u denotes the Riesz measure corresponding to u. It follows from the Riesz decomposition theorem that a potential on W is defined by

$$w(X) = \sum_k a_k^{-1} G_W(X_k, X) \qquad (X \in W),$$

where G_W denotes the Green function for W. Let $T = \cup_k V_k$. Then the restriction of w to $\partial T \cap W$ is $\mu_{T,X}$-integrable when $X \in T$. By monotone convergence

$$\int_{\partial T \cap W} w(Y) d\mu_{T,X}(Y) = \sum_k a_k^{-1} \int_{\partial T \cap W} G_W(X_k, Y)\, d\mu_{T,X}(Y)$$
$$= \sum_k a_k^{-1} \int_{\partial T \cap W} G_W(X, Y)\, d\mu_{T,X_k}(Y).$$

(The second equality follows from the symmetry of the Green functions for the sets W and T.) However, (3.5) yields

$$\sum_k a_k^{-1} \mu_{T,X_k}(U_0 \cap \partial T) = \sum_k a_k^{-1} \mu_{V_k,X_k}(U_0 \cap \partial V_k) = +\infty.$$

Hence the Riesz measure associated with the superharmonic function

$$X \mapsto \int_{\partial T \cap W} w\, d\mu_{T,X}$$

is infinite on the compact set $\overline{U}_0 \cap \partial T$, which is impossible. The theorem is now proved.

3.8 Runge Approximation

As a result of the material in Chapter 2 and the above sections of this chapter we can now state a generalization of Theorem 1.10 which deals with approximation on unbounded sets. It not only characterizes the pairs (Ω, E) for which functions in $\mathcal{H}(E)$ can be uniformly approximated (on E) by functions in $\mathcal{H}(\Omega)$, but also shows that a stronger form of approximation is generally possible. As in the preceding few sections, Ω is a connected open set in $\mathbf{R}^n$, and E is a relatively closed proper subset of Ω.

Theorem 3.15. *The following are equivalent:*
(a) for each u in $\mathcal{H}(E)$ and each positive number ϵ there exists v in $\mathcal{H}(\Omega)$ such that $|v - u| < \epsilon$ on E;
(b) for each u in $\mathcal{H}(E)$ and each s in $\mathcal{S}^+(E)$ there exists v in $\mathcal{H}(\Omega)$ such that $0 < v - u < s$ on E;
(c) $\Omega \backslash \widehat{E}$ and $\Omega \backslash E$ are thin at the same points of E and (Ω, E) satisfies the (K, L)-condition.

Proof. Clearly (b) implies (a). Further, it follows from Theorems 3.14 and 1.14 that (a) implies (c). It remains to show that (c) implies (b). We will do this by first establishing that (c) implies the following apparently weaker form of condition (b):

(b′) for each u in $\mathcal{H}(E)$ and each s in $\mathcal{S}^+(\widehat{E})$, there exists v in $\mathcal{H}(\Omega)$ such that $0 < v - u < s$ on E.

Suppose that (c) holds, let $u \in \mathcal{H}(E)$ and $s \in \mathcal{S}^+(\widehat{E})$. We first observe, using the reasoning given for Lemmas 1.11 and 1.12, that each component V of $\widehat{E} \backslash E$ is regular for the Dirichlet problem and satisfies $\partial V \subseteq \partial \widehat{E}$. Also, we know that $\widehat{E} \neq \Omega$, for otherwise (by (c)) $E = \Omega$. Next, as in the first paragraph of §3.6, we may assume that $s \in \mathcal{H}^+(\widehat{E})$, and it is enough to show that there exists v in $\mathcal{H}(\Omega)$ such that $|v - u| < s$ on E. There are open subsets ω_1 and ω_2 of Ω such that u is harmonic on ω_1 and $E \subset \omega_1$, and s is positive and harmonic on ω_2 and $\widehat{E} \subset \omega_2$. We may assume ω_2 to be Greenian. We denote by $\{V_k : k \in I\}$, where $I \subseteq \mathbf{N}$, the collection of Ω-bounded components of $\Omega \backslash E$ which are not subsets of ω_1, choose Y_k in V_k for each k in I, and define

$$\omega_3 = \omega_1 \cup \left(\bigcup_{k \in I} (V_k \backslash \{Y_k\}) \right).$$

Let K be an arbitrary compact subset of Ω, and define

$$S = \cup_{k \in J} V_k, \text{ where } J = \{k \in I : \overline{V_k} \cap K \neq \emptyset\}.$$

It follows from the (K, L)-condition that S is Ω-bounded. Since

$$\operatorname{dist}(\overline{S} \cap E, \mathbf{R}^n \backslash \omega_1) > 0,$$

the set J must be finite. Thus $\{Y_k : k \in I\}$ has no limit point in Ω. It also follows that we can choose a collection $\{U_k : k \in I\}$ of Ω-bounded open sets such that $\overline{V_k} \subset U_k$ and $\overline{U_k} \subset \omega_2$ for each k, and such that any given compact subset of Ω intersects only finitely many of the sets $\overline{U_k}$.

Next, for each V_k, we follow the construction in the second paragraph of §1.8. This results in a superharmonic function u' on ω_3 such that $u' = u$ on an open set which contains E and such that, for each k in I, the function $u'(X) + b_k \phi_n(|X - Y_k|)$ has a superharmonic extension to $\omega_3 \cup \{Y_k\}$ for a suitable choice of positive constant b_k.

Now fix k temporarily. For each m in $\mathbf{N}$ we define the set

$$A_{k,m} = \{X \in \overline{U_k} : \operatorname{dist}(X, \widehat{E}) \geq 1/m\},$$

and let $g_{k,m}$ be the Green function for $\omega_2 \backslash A_{k,m}$ with pole at Y_k. Arguing as in the fourth paragraph of §1.8, we see that $g_{k,m} \downarrow g_k$ on V_k as $m \to \infty$, where g_k is the Green function for V_k with pole at Y_k. Further, since V_k is regular and s has a positive lower bound on $\overline{V_k}$, there is a compact subset K_k of V_k such that $Y_k \in K_k^\circ$ and $g_k < 2^{-k-2} b_k^{-1} s$ on $V_k \backslash K_k^\circ$. Hence, by the monotonicity of the sequence $(g_{k,m})_{m \geq 1}$ and Dini's Theorem, there exists m_k such that

$$g_{k,m_k}(X) \leq g_k(X) + 2^{-k-2} b_k^{-1} s(X) < 2^{-k-1} b_k^{-1} s(X) \qquad (X \in \partial K_k).$$

It follows that $g_{k,m_k}(X) \leq 2^{-k-1} b_k^{-1} s(X)$ on $\omega_2 \backslash A_{k,m_k}$, and hence on E.

We now define

$$W = \omega_2 \backslash \left(\bigcup_{k \in I} A_{k,m_k} \right).$$

This is an open set because only finitely many of the sets A_{k,m_k} intersect a given compact subset of Ω. Also, $\widehat{E} \subset W$. Using $G_W(\cdot\,,\cdot)$ to denote the Green function for W, we have

$$G_W(Y_k, X) \leq g_{k,m_k}(X) \leq 2^{-k-1} b_k^{-1} s(X) \qquad (X \in E).$$

It follows that the equation

$$v_1(X) = \sum_{k \in I} b_k G_W(Y_k, X) \qquad (X \in W)$$

defines a potential on W, and that $v_1 \leq s/2$ on E. The function $v_2 = u' + v_1$, suitably redefined on the set where it is the difference of two infinite values, is superharmonic on the open set $W \cap (\omega_1 \cup (\cup_k V_k))$, which contains $\widehat{E}$. Also, $u \leq v_2 \leq u + s/2$ on E. Thus, if we define $v_3 = v_2 + s/4$, we obtain a function in $\mathcal{S}(\widehat{E})$ such that $u < v_3 \leq u + 3s/4$ on E.

Replacing u by $-u$ in the above argument, we obtain v_4 in $\mathcal{S}(\widehat{E})$ such that $-u < v_4 \leq -u + 3s/4$ on E, whence $-v_4 < v_3$ on E. It follows from the maximum principle that $-v_4 < v_3$ on an open set ω which contains $\widehat{E}$. The greatest harmonic minorant w of v_3 on ω therefore satisfies $-v_4 \leq w \leq v_3$ on E, whence $|w - u| \leq 3s/4$ on E. We observe that $w \in \mathcal{H}(\widehat{E})$. Further, $\Omega^* \backslash \widehat{E}$ is connected (obviously) and also locally connected (by the (K, L)-condition). Thus we may apply Theorem 3.11 to obtain v in $\mathcal{H}(\Omega)$ such that $0 < v - w < s/4$ on $\widehat{E}$. Hence $|v - u| < s$ on E, proving (b′) (see the observations made in the second paragraph of this proof).

The proof of (b) (and hence of Theorem 3.15) will be completed by proving the following lemma.

Lemma 3.16. *If condition (c) of Theorem 3.15 holds and $s \in \mathcal{S}^+(E)$, then there exists w in $\mathcal{S}^+(\widehat{E})$ such that $w < s$ on E.*

Proof of Lemma. Suppose that (c) holds and let $s \in \mathcal{S}^+(E)$. As above, we may assume that $s \in \mathcal{H}^+(E)$. In view of the (K, L)-condition we may choose an exhaustion (K_m) of Ω such that any Ω-bounded component of $\Omega \backslash E$ which intersects K_m is contained in K_{m+1}. We can also arrange that $K_1 \cap \widehat{E} = \emptyset$. We now define

$$\epsilon_m = \inf\left(\left\{s(X) : X \in E \cap K_m\right\} \cup \{1\}\right) \qquad (m \in \mathbf{N}). \tag{3.6}$$

For each m we use the fact that (b′) holds (with s replaced by a positive constant) to obtain w_m in $\mathcal{H}(\Omega)$ such that

$$\left(1 - \frac{1}{m}\right) s(X) < w_m(X) < \left(1 - \frac{1}{m}\right) s(X) + \frac{\epsilon_m}{m(m+1)} \quad (X \in E). \tag{3.7}$$

If $X \in E \cap K_m$, then

$$\begin{aligned} w_{m+1}(X) &> \left(1 - \frac{1}{m+1}\right) s(X) \\ &\geq \left(1 - \frac{1}{m}\right) s(X) + \frac{\epsilon_m}{m(m+1)} > w_m(X). \end{aligned} \tag{3.8}$$

It follows from the minimum principle and our construction of (K_m) that $w_{m+1} > w_m$ on $\widehat{E} \cap K_{m-1}$. It also follows from the minimum principle that $w_m > 0$ on an open set U_m which contains $\widehat{E}$. Let V_m be the open subset of $U_m \cap U_{m+1}$ where $w_{m+1} > w_m$, so that $\widehat{E} \cap K_{m-1} \subset V_m$, and define

$$W_m = (K^\circ_{m+2} \backslash K_m) \cap \left(\bigcap_{k=1}^{m+2} U_k\right) \cap V_{m+3} \qquad (m \geq 1).$$

Thus W_m is an open set which contains $\widehat{E} \cap (K^\circ_{m+2} \backslash K_m)$. It follows that the set W, given by $W = \cup_m W_m$, is open and contains $\widehat{E}$. We define a function on W by

$$w(X) = \min\{w_1(X), w_2(X), \ldots, w_{m+3}(X)\} \qquad (X \in W_m; m \in \mathbf{N}).$$

To see that w is well-defined, we note that $w_{m+4} > w_{m+3}$ on V_{m+3} by the definition of V_m, so

$$\min\{w_1(X), \ldots, w_{m+4}(X)\} = \min\{w_1(X), \ldots, w_{m+3}(X)\} \qquad (X \in W_m \cap W_{m+1}).$$

Clearly $w \in \mathcal{S}^+(W)$. Finally, if $X \in E \cap (K^\circ_{m+2} \backslash K_m)$, then it follows from (3.7) that

$$w(X) \leq w_{m+3}(X) < \left(1 - \frac{1}{m+4}\right) s(X) < s(X),$$

whence $w < s$ on E, as required.

3.9 Approximation by Functions in $\mathcal{H}(E)$

The next result is a generalization of Theorem 1.3 which deals with approximation by functions harmonic on a neighbourhood of a closed set.

Theorem 3.17. *The following are equivalent:*
(a) for each u in $C(E) \cap \mathcal{H}(E^\circ)$ and each positive number ϵ there exists v in $\mathcal{H}(E)$ such that $|v - u| < \epsilon$ on E;
(b) for each u and s in $C(E) \cap \mathcal{H}(E^\circ)$, where $s > 0$, there exists v in $\mathcal{H}(E)$ such that $0 < v - u < s$ on E;
(c) $\Omega \backslash E$ and $\Omega \backslash E^\circ$ are thin at the same points of E.

We remark that Theorem 3.17 remains true if, in (b), we require only that the positive function s belong to $C(E) \cap \mathcal{S}(E^\circ)$. This will follow from Lemma 6.12.

Proof. Clearly (b) implies (a). Further, it follows from Theorem 1.6 that (a) implies (c). It remains to show that (c) implies (b).

In fact, we will temporarily assume that Ω has a Green function $G(\cdot\,,\cdot)$, and use the fusion result in Chapter 2 to see that (c) implies the following apparently weaker form of (b):

(b′) for each u in $C(E) \cap \mathcal{H}(E^\circ)$ and each positive number ϵ there exists v in $\mathcal{H}(E)$ such that $|v(X) - u(X)| < \epsilon g(X)$ on E.

Here, as before, $g(X) = \min\{1, G(Q, X)\}$, where Q is some fixed point of Ω.

Suppose that (c) holds, let $u \in C(E) \cap \mathcal{H}(E^\circ)$ and $\epsilon > 0$. Also, let (K_m) be an exhaustion of Ω such that, for each m, the set $\Omega \backslash K_m$ is not thin at any point of ∂K_m. We define $F_m = K_m \cap E$ for each m, and observe from condition (c) that $\Omega \backslash F_m$ and $\Omega \backslash F^\circ_m$ are thin at the same points. It follows from Theorem 1.3 and Lemma 1.8 that, for any ϵ and m, there exists q in $\mathcal{I}(\mathbf{R}^n)$ such that

$$\left|(q-u)(X)\right| < \epsilon g(X) \qquad (X \in F_m).$$

With this extra ingredient the proof of (b′) is identical to that of Theorem 3.5.

We next revert to our original assumption that Ω is any connected open set in $\mathbf{R}^n$, and use what we have proved above to establish the following lemma.

Lemma 3.18. *If (c) holds and $s \in C(E) \cap \mathcal{H}(E^\circ)$, where $s > 0$, then there exists w in $\mathcal{H}(E)$ such that $0 < w < s$ on E.*

Proof of Lemma. By deleting a closed ball from $\Omega \backslash E$ if necessary, we may assume that Ω is Greenian. Let (K_m) be an exhaustion of Ω such that $K_1 \cap E = \emptyset$. We define ϵ_m as in (3.6). For each m we use the fact that (c) implies (b′) to obtain w_m in $\mathcal{H}(E)$ such that (3.7) holds. Thus, if $X \in E \cap K_m$, then (3.8) holds. Let U_m be the open set where w_m is defined and positive, and let V_m be the open subset of $U_m \cap U_{m+1}$ where $w_{m+1} > w_m$. Thus $E \cap K_m \subset V_m$. Let

$$W_m = (K_{m+2}^\circ \backslash K_m) \cap \left(\bigcap_{k=1}^{m+1} U_k \right) \cap V_{m+2} \qquad (m \geq 1).$$

Thus W_m is an open set which contains $E \cap (K_{m+2}^\circ \backslash K_m)$. It follows that the set W, given by $W = \cup_m W_m$, is open and contains E. We define a function on W by

$$w'(X) = \min\{w_1(X), w_2(X), \ldots, w_{m+2}(X)\} \qquad (X \in W_m; m \in \mathbf{N})$$

and observe, as in the proof of Lemma 3.16, that w' is well-defined and satisfies $w' < s$ on E. Clearly $w' \in \mathcal{S}^+(W)$. In each component ω of W we replace w' by its reduced function relative to a closed ball contained in $\omega \backslash E$. The resulting function w belongs to $\mathcal{H}^+(E)$ and also satisfies $w < s$ on E. The lemma is proved.

We now complete the proof of Theorem 3.17. Suppose that (c) holds, let $u, s \in C(E) \cap \mathcal{H}(E^\circ)$ and suppose that $s > 0$. By Lemma 3.18 there exists a positive harmonic function w on an open set ω which contains E, such that $w < s$ on E. We can assume that ω is Greenian. Let ω_0 be any component of ω, let B be a closed ball in ω_0 with centre Q, and choose a in the interval $[1, +\infty)$ such that

$$a > \sup\{G_0(Q, X) : X \in \omega_0 \backslash B\},$$

where G_0 is the Green function for ω_0. If we define

$$b = \inf\left\{\frac{w(X)}{\min\{a, G_0(Q,X)\}} : X \in B\right\},$$

then

$$w(X) \geq b\min\{a, G_0(Q,X)\} \geq b\min\{1, G_0(Q,X)\} \qquad (X \in \Omega).$$

It follows from the fact that (c) implies (b′) that there exists v in $\mathcal{H}(E \cap \omega_0)$ such that $|v - (u + w/2)| < w/2$, whence $0 < v - u < w < s$ on $E \cap \omega_0$. Repeating this argument for each component of ω, we obtain v in $\mathcal{H}(E)$ such that $0 < v - u < s$ on E.

3.10 Arakelyan Approximation

Next we present a generalization of Theorem 1.15 which includes the harmonic analogue of Arakelyan's Theorem (see §3.1).

Theorem 3.19. *The following are equivalent:*
(a) for each u in $C(E) \cap \mathcal{H}(E^\circ)$ and each positive number ϵ there exists v in $\mathcal{H}(\Omega)$ such that $|v - u| < \epsilon$ on E;
(b) for each u and s in $C(E) \cap \mathcal{H}(E^\circ)$, where $s > 0$, there exists v in $\mathcal{H}(\Omega)$ such that $0 < v - u < s$ on E;
(c) $\Omega \backslash \widehat{E}$ and $\Omega \backslash E^\circ$ are thin at the same points of E and (Ω, E) satisfies the (K, L)-condition.

As was the case with Theorem 3.17, the results of Chapter 6 show that the positive function s in (b) need only belong to $C(E) \cap \mathcal{S}(E^\circ)$. Before proving Theorem 3.19 we present an example of its application.

Example 3.20. Let $\Omega = \mathbf{R}^n$ and let E be a closed subset of $\mathbf{R}^{n-1} \times [0, +\infty)$ such that condition (c) of Theorem 3.19 holds. Then, for each u in $C(E) \cap \mathcal{H}(E^\circ)$, each positive number ϵ, and each continuous function $f : \mathbf{R}^{n-1} \to (0, 1]$, there exists v in $\mathcal{H}(\mathbf{R}^n)$ such that

$$0 < v(X) - u(X) < \epsilon\left(\frac{1 + x_n}{1 + |X|^n}\right) \qquad (X \in E; x_n \geq 0)$$

and

$$0 < v(X) - u(X) < f(x_1, \ldots, x_{n-1}) \qquad (X \in E; x_n = 0).$$

Thus the error in the approximation can be arranged to decay arbitrarily quickly on the hyperplane boundary.

Details of Example. Let $\omega = \mathbf{R}^{n-1} \times (0, +\infty)$, let $\epsilon > 0$ and let $f : \mathbf{R}^{n-1} \to (0, 1]$ be continuous. Further, let h be the Poisson kernel for $\mathbf{R}^{n-1} \times (-1, +\infty)$ defined by

$$h(X', x_n) = \frac{1 + x_n}{\{|X'|^2 + (1 + x_n)^2\}^{n/2}} \qquad (X = (X', x_n) \in \mathbf{R}^{n-1} \times \mathbf{R}),$$

let $f_1(X', 0) = \min\{f(X'), h(X', 0)\}$, and define s to be equal to the half-space Poisson integral of f_1 in ω and equal to f_1 on $\partial\omega$. Thus $s \in C(\overline{\omega}) \cap \mathcal{H}(\omega)$ and $s > 0$. It is clear from the maximum principle that $s \leq h$ on ω. The assertions of Example 3.20 now follow immediately from Theorem 3.19.

Proof of Theorem 3.19. Clearly (b) implies (a). It follows from Theorems 3.15 and 3.17 that (a) implies (c). Now suppose that (c) holds, let $u, s \in C(E) \cap \mathcal{H}(E^\circ)$ and suppose that $s > 0$. Since $\Omega \backslash \widehat{E} \subset \Omega \backslash E \subset \Omega \backslash E^\circ$, we can apply Lemma 3.18 to obtain w in $\mathcal{H}(E)$ such that $0 < w < s$ on E. It now follows from Theorems 3.15 and 3.17 that (b) holds.

As with approximation on compact sets things simplify when $n = 2$. The following result may be obtained by an argument similar to that presented for Corollary 1.16.

Corollary 3.21. *Let $n = 2$. The following are equivalent:*
(a) for each u in $\mathcal{H}(E)$ and each positive number ϵ there exists v in $\mathcal{H}(\Omega)$ such that $|v - u| < \epsilon$ on E;
(b) for each u and s in $C(E) \cap \mathcal{H}(E^\circ)$, where $s > 0$, there exists v in $\mathcal{H}(\Omega)$ such that $0 < v - u < s$ on E;
(c) $\partial\widehat{E} = \partial E$ and (Ω, E) satisfies the (K, L)-condition.

We note, in particular, that condition (c) above is satisfied if $\Omega^* \backslash E$ is connected and locally connected. (This is not true of condition (c) of Theorem 3.19 when $n \geq 3$.)

3.11 Weak Approximation

Finally, in this chapter, we investigate which pairs (Ω, E) have the property that, for any u in $\mathcal{H}(E)$, there exists v in $\mathcal{H}(\Omega)$ such that $v - u$ is bounded on E.

Theorem 3.22. *Suppose that Ω is Greenian. The following are equivalent:*
(a) for each u in $\mathcal{H}(E)$ there exists v in $\mathcal{H}(\Omega)$ and a positive number a such

that $|v-u|<a$ on E;
(b) there is a compact subset J of Ω such that $\Omega\backslash\widehat{E}$ and $\Omega\backslash E$ are thin at the same points of $E\backslash J$, and (Ω, E) satisfies the (K,L)-condition.

We remark that the theorem becomes false if we drop the assumption that Ω is Greenian. To see this, suppose that Ω is not Greenian (and hence $n=2$), let $B(Y,r)$ be a ball whose closure is contained in Ω, let $E=\Omega\backslash B(Y,r)$ and let $u(X)=\phi_2(|X-Y|)$. Then (b) holds, but (a) cannot hold, for otherwise $u-v+a$ would be a non-constant positive superharmonic function on Ω, and this yields a contradiction.

Proof. Suppose that Ω is Greenian and that (b) holds. Let K_1 be a compact subset of Ω such that $J\subset K_1^\circ$, and let u be a harmonic function on an open set ω such that $E\subset\omega\subseteq\Omega$. It follows from the (K,L)-condition that there are only finitely many Ω-bounded components $V_1,V_2,\ldots,V_m$ of $\Omega\backslash E$ which satisfy both $\overline{V_k}\cap K_1\neq\emptyset$ and $\overline{V_k\backslash\omega}\neq\emptyset$. For each k in $\{1,2,\ldots,m\}$ we choose X_k and r_k such that $B(X_k,r_k)\subset V_k$. Next we define $\Omega_1=\Omega\backslash\{X_1,X_2,\ldots,X_m\}$ and define E_1 to be the union of E with the Ω-bounded components of $\Omega\backslash E$ which are contained in ω. The pair (Ω_1,E_1) satisfies condition (c) of Theorem 3.15 so there exists w in $\mathcal{H}(\Omega_1)$ such that $|w-u|<1$ on E_1. Using the Laurent expansions for w about the points $X_1,X_2,\ldots,X_m$, we can write

$$w=v+w_1+w_2+\ldots+w_m,$$

where $v\in\mathcal{H}(\Omega)$, and $w_k\in\mathcal{H}(\mathbf{R}^n\backslash\{X_k\})$ for each k. When $n\geq 3$ we have

$$\big|w_k(X)\big|\leq a_k \qquad \big(X\in\mathbf{R}^n\backslash B(X_k,r_k);k\in\{1,2,\ldots,m\}\big), \tag{3.9}$$

where a_k is some positive constant. When $n=2$ we can replace the term $\phi_2(|X-X_k|)$ in the Laurent expansion of w about X_k by the term $G_\Omega(X_k,X)$, where G_Ω is the Green function for Ω, and then (3.9) again holds. It thus follows that $v-u$ is bounded on E, as required.

Conversely, suppose that (a) holds. It follows from Theorem 3.14 that the (K,L)-condition holds. Next, let (V_k) be a sequence of Ω-bounded components of $\Omega\backslash E$ such that only a finite number of the sets V_k intersect any given compact subset of Ω. (If no such sequence exists, then the thinness condition in (b) must hold, in view of the (K,L)-condition.) For each k we fix X_k in V_k, and define

$$w_k(X)=G_\Omega(X_k,X)-R^{\Omega\backslash\widehat{E}}_{G_\Omega(X_k,\,\cdot\,)}(X) \qquad (X\in\Omega).$$

We choose, if possible, a point Y_k in E such that $w_k(Y_k)>0$, and then define $b_k=k/w_k(Y_k)$. If $w_k=0$ on all of E, then we define $b_k=1$. Next we apply Corollary 3.9 to obtain a harmonic function u on the set $\Omega_2 = \Omega\backslash\{X_k : k \in \mathbf{N}\}$ such that, for each k, the function $u(X) + b_k\phi_n(|X - X_k|)$ has a harmonic extension to $\Omega_2 \cup \{X_k\}$. By hypothesis there is a harmonic function v on Ω and a positive number a such that $|v - u| < a$ on E. We define the open set

$$W = \{X \in \Omega : v(X) - u(X) + a > 0\},$$

and observe from the minimum principle that $\widehat{E} \subseteq W$. Also,

$$v(X) - u(X) + a \geq b_k G_W(X_k, X) \qquad (X \in W; k \in \mathbf{N}),$$

where G_W denotes the Green function for W. Hence $b_k G_W(X_k, \cdot) < 2a$ on E, and so

$$w_k(X) \leq G_\Omega(X_k, X) - R^{\Omega\backslash W}_{G_\Omega(X_k,\cdot)}(X) = G_W(X_k, X) < 2a/b_k \qquad (X \in E).$$

Thus $w_k(Y_k) < 2a w_k(Y_k)/k$ for each k such that Y_k exists as described above, and so we must have

$$G_\Omega(X_k, X) = R^{\Omega\backslash\widehat{E}}_{G_\Omega(X_k,\,\cdot\,)}(X) = R^{\Omega\backslash\widehat{E}}_{G_\Omega(X,\,\cdot\,)}(X_k) \qquad (X \in E; k > k')$$

for some choice of k'. It follows from the maximum principle that

$$G_\Omega(Y, X) = R^{\Omega\backslash\widehat{E}}_{G_\Omega(X,\,\cdot\,)}(Y) \qquad (X \in E; Y \in V_k; k > k').$$

In fact, in view of the arbitrary choice of the sequence (V_k), there must be a compact subset K of Ω such that

$$G_\Omega(Y, X) = R^{\Omega\backslash\widehat{E}}_{G_\Omega(X,\,\cdot\,)}(Y) \qquad (X \in E; Y \in V),$$

for every Ω-bounded component V of $\Omega\backslash E$ which does not intersect K. If we define E' to be the union of E with all other Ω-bounded components of $\Omega\backslash E$, then

$$R^{\Omega\backslash\widehat{E}}_{G_\Omega(X,\,\cdot\,)} = R^{\Omega\backslash E'}_{G_\Omega(X,\cdot)} \qquad (X \in E).$$

Thus $\Omega\backslash\widehat{E}$ and $\Omega\backslash E'$ are thin at the same points of E, by Theorem 0.J. It follows from the (K, L)-condition that there is a compact subset J of Ω such that $E\backslash J = E'\backslash J$, and so the thinness condition in (b) holds.

Notes

An early result of the type considered in this chapter is due to Shaginyan [Sha1], who proved Theorem 3.19 in the special case where $E^\circ = \emptyset$. Gauthier, Goldstein and Ow proved Corollary 3.8 when $n = 2$ in [GGO1], and when $n \geq 3$ in [GGO2], and showed that all partial derivatives of u, up to a predetermined order, can be simultaneously approximated by the corresponding partial derivatives of v. Corollary 3.9 also occurs in [GGO2]. The equivalence of (a) and (c) in Theorem 3.17 is due to Labrèche [Lab] (see also [BB] and [PV] for recent generalizations). These results were subsequently refined by Armitage and Goldstein [AG2], who proved Theorem 3.5, Corollary 3.6 and the implication "(c) $\Rightarrow$ (b$'$)" in §3.9 above. Theorems 3.11, 3.14, 3.15, 3.17 $\big($(c) $\Rightarrow$ (b)$\big)$, 3.19 and 3.22 are due to the author: see [Gar3], [Gar5] and [Gar6]. In connection with Theorem 3.14, Bagby and Gauthier [BG2] had earlier shown that Runge approximation requires $\Omega^*\backslash\widehat{E}$ to be locally connected. Also, in relation to Theorem 3.19, mention should be made of [Sha2], [GO] and [BG2]; to the best of our knowledge no proofs of the results announced in [Sha2] have yet appeared. For holomorphic results related to Theorem 3.22 the reader is referred to [Ner2] and [GHS]. Finally, the papers [GGO1], [GGO2] identify the pairs (Ω, W), where W is an open subset of Ω, with the following property: for each u in $\mathcal{H}(W)$, each subset F of W which is relatively closed in Ω and each positive number ϵ, there exists v in $\mathcal{H}(\Omega)$ such that $|v - u| < \epsilon$ on F. It is shown that (Ω, W) has this property if and only if $\Omega^*\backslash W$ is connected.

4 Carleman Approximation

4.1 Introduction

In 1927 Carleman [CarT] proved the following remarkable generalization of the classical Weierstrass approximation theorem: given any continuous function $f : \mathbf{R} \to \mathbf{C}$ and any continuous "error" function $\epsilon : \mathbf{R} \to (0, 1]$, there is an entire (holomorphic) function g such that $|g - f| < \epsilon$ on $\mathbf{R}$. Thus the function $\epsilon(x)$ can be chosen to decay arbitrarily quickly as $|x| \to \infty$. The problem of identifying which pairs (Ω, E) permit Carleman-type holomorphic approximation was solved by Nersesyan [Ner1] in the following result.

Nersesyan's Theorem (1971). *Let Ω be a connected open set in $\mathbf{C}$ and E be a relatively closed proper subset of Ω. The following are equivalent:*
(a) for every f in $C(E) \cap \mathrm{Hol}(E^\circ)$ and every continuous function $\epsilon : E \to (0, 1]$ there exists g in $\mathrm{Hol}(\Omega)$ such that $|g - f| < \epsilon$ on E;
(b)(i) $\Omega^* \backslash E$ is connected and locally connected, and
(ii) for each compact subset K of Ω there is a compact subset L of Ω which contains every component of E° that intersects K.

It was Gauthier [Gau1] who first realised the significance of the restriction on the components of E° in (b)(ii) above. We will discuss this condition in §4.3. This chapter presents several results on harmonic approximation with arbitrary error function, including a direct analogue of Nersesyan's Theorem.

Throughout this chapter Ω denotes a connected open set in $\mathbf{R}^n$ and E is a relatively closed proper subset of Ω.

4.2 Decay of Harmonic Functions

The purpose of this section is to prove the following result, which shows that there is a limit to how quickly a non-constant harmonic function on a given unbounded connected open set can decay near infinity.

Theorem 4.1. *Let ω be an unbounded connected open set in $\mathbf{R}^n$. Then there exists a continuous function $\epsilon_\omega : [0,+\infty) \to (0,1]$ with the following property: if $u \in \mathcal{H}(\omega)$ and $|u(X)| \leq \epsilon_\omega(|X|)$ on ω, then $u \equiv 0$.*

The following simple lemma will be used in the proof of Theorem 4.1.

Lemma 4.2. *Let ω be a connected open set, let K be a compact subset of ω and W be a non-empty open subset of ω. Then, for each positive number η there exists a positive number δ with the following property: if $u \in \mathcal{H}(\omega)$ and $|u| \leq 1$ on ω and $|u| \leq \delta$ on W, then $|u| \leq \eta$ on K.*

Proof of Lemma. Let ω, K, W be as in the statement of the lemma, but suppose that the conclusion of the lemma is false. Then there exists a positive number η and a sequence (u_m) of harmonic functions on ω such that $|u_m| \leq 1$ on ω and $|u_m| \leq m^{-1}$ on W, yet $\sup_K |u_m| > \eta$. Since (u_m) is uniformly bounded on ω, there is a subsequence which converges locally uniformly on ω to some harmonic function u. Clearly $u = 0$ on W, whence $u \equiv 0$ on ω. However, this contradicts the fact that $\sup_K |u_m| > \eta$ for each m.

Proof of Theorem 4.1. Let ω be an unbounded connected open set in $\mathbf{R}^n$, let K be a compact subset of ω with non-empty interior, and let (X_m) be a sequence of points in ω such that $|X_m| \to \infty$. For each m, let δ_m be the positive number δ in Lemma 4.2 corresponding to $W = B(X_m, 1) \cap \omega$ and $\eta = m^{-1}$. Next, let $\epsilon : [0, +\infty) \to (0,1]$ be a continuous function such that $\epsilon(|X|) \leq \delta_m$ on $B(X_m, 1)$, for each m. Now let u be a harmonic function on ω such that $|u(X)| \leq \epsilon(|X|)$ on all of ω. Since $|u| \leq \delta_m$ on $B(X_m, 1) \cap \omega$, we see from Lemma 4.2 that $|u| \leq m^{-1}$ on K. Since this holds for all m, and $K^\circ \neq \emptyset$, it follows that $u \equiv 0$.

4.3 Approximation by Functions in $\mathcal{H}(E)$

We will say that (Ω, E) satisfies the *long islands condition* if, for each compact subset K of Ω, there is a compact subset L of Ω which contains every component of E° that intersects K. This condition is trivially satisfied if $E^\circ = \emptyset$. Also, it clearly implies that every component of E° is Ω-bounded. The following example shows that it implies more than this. It also motivates the terminology.

Example 4.3. Let

$$E = \Big(\{0\} \times [0, +\infty)\Big) \cup \left(\bigcup_{k=1}^{\infty}\Big(\big[(2k+1)^{-1}, (2k)^{-1}\big] \times [0, k]\Big)\right).$$

Then $(\mathbf{R}^2, E)$ does not satisfy the long islands condition, as can be seen by choosing K to be $[0,1]^2$.

Our first main result characterizes those pairs (Ω, E) which permit functions in $C(E) \cap \mathcal{H}(E^\circ)$ to be approximated by functions in $\mathcal{H}(E)$ with arbitrary error function.

Theorem 4.4. *The following are equivalent:*
(a) for each u in $C(E) \cap \mathcal{H}(E^\circ)$ and each continuous function $\epsilon : E \to (0, 1]$, there exists v in $\mathcal{H}(E)$ such that $|v - u| < \epsilon$ on E;
(b) for each u in $C(E) \cap \mathcal{H}(E^\circ)$ and each continuous function $\epsilon : E \to (0, 1]$, there exists v in $\mathcal{H}(E)$ such that $0 < v - u < \epsilon$ on E;
(c) $\Omega \backslash E$ and $\Omega \backslash E^\circ$ are thin at the same points of E, and (Ω, E) satisfies the long islands condition.

In order to prove that (c) implies (b) in Theorem 4.4 we will require the following lemma.

Lemma 4.5. *If the pair (Ω, E) satisfies condition (c) of Theorem 4.4 and $\epsilon : E \to (0, 1]$ is continuous, then there exists s in $\mathcal{S}^+(E)$ such that $s < \epsilon$ on E.*

Proof of Lemma. We may assume, by deleting from Ω a closed ball contained in $\Omega \backslash E$ if necessary, that Ω is Greenian. Using condition (c) of Theorem 4.4 we can construct an exhaustion (K_m) of Ω such that $K_1^\circ \neq \emptyset$ and such that every component V of E° which satisfies $\overline{V} \cap K_m \neq \emptyset$ also satisfies $\overline{V} \subset K_{m+1}^\circ$.

Next we define $A[l, m] = K_m \backslash K_l^\circ$ and $A(l, m) = K_m^\circ \backslash K_l$ whenever $l < m$, and also

$$F(m;k) = \{X \in A[m,m+5] : \operatorname{dist}(X,E) \geq k^{-1}\} \qquad (m,k \in \mathbf{N})$$

and

$$\delta_m = \inf\Big(\{\epsilon(X) : X \in E \cap A[m,m+1]\} \cup \{1\}\Big) \qquad (m \in \mathbf{N}).$$

Let $Q \in K_1^\circ$ and let g' denote the Green function for Ω with pole at Q. Then

$$\widehat{R}_{g'}^{F(m;k)}(X) \uparrow \widehat{R}_{g'}^{A[m,m+5]\backslash E}(X) \qquad (k \to \infty; X \in \Omega; m \in \mathbf{N}).$$

If $u_1 \in \mathcal{S}^+(\Omega)$ and $u_1(X) \geq g'(X)$ on $A[m,m+5]\backslash E$, then the same inequality holds for points X of the set

$$\{X \in E : A[m,m+5]\backslash E \text{ is not thin at } X\}.$$

Since condition (c) of Theorem 4.4 holds, we know that $\Omega\backslash E$ and $\Omega\backslash E^\circ$ are thin at the same points of E. Further, the set of points of ∂E where $\Omega\backslash E^\circ$ is thin is a polar set, by Theorem 0.E. It follows that

$$\widehat{R}_{g'}^{F(m;k)}(X) \uparrow \widehat{R}_{g'}^{S}(X) \qquad (k \to \infty; X \in \Omega),$$

where

$$S = (A[m,m+5]\backslash E) \cup (A(m,m+5)\backslash E^\circ).$$

The construction of the sequence (K_m) and the minimum principle together show that, if $u_2 \in \mathcal{S}^+(\Omega)$ and $u_2 \geq g'$ on S, then $u_2 \geq g'$ on $A[m+1,m+4]$, whence $\widehat{R}_{g'}^{S} = g'$ on $A[m+2,m+3]$. Thus, by Dini's Theorem, there exists k_m such that

$$g'(X) \geq \widehat{R}_{g'}^{F(m;k_m)}(X) > g'(X) - \delta_{m+2} \qquad (X \in A[m+2,m+3]). \quad (4.1)$$

We now define the set

$$F_1 = \bigcup_{m=1}^{\infty} F(m;k_m).$$

Clearly F_1 is a relatively closed subset of Ω such that $K_1^\circ \cap F_1 = \emptyset$ and $E \subset \Omega\backslash F_1$. We also define the positive number

$$\delta_0 = \inf\Big(\{\epsilon(X) : X \in E \cap K_3\} \cup \{1\}\Big)$$

and the function

$$s_1(X) = 2^{-1}\min\Big\{g'(X) - \widehat{R}_{g'}^{F_1}(X), \delta_0/2\Big\} \quad (X \in \Omega\backslash F_1). \qquad (4.2)$$

We observe that s_1 is non-negative and superharmonic on $\Omega\backslash F_1$ and positive on the component of K_1° which contains Q. Also, since

$$\widehat{R}_{g'}^{F_1}(X) \geq \widehat{R}_{g'}^{F(m;k_m)}(X) \qquad (X \in \Omega; m \in \mathbf{N}),$$

it follows from (4.1) and (4.2) that $s_1(X) < 2^{-1}\epsilon(X)$ on E.

The above argument can be repeated, with the sequence $(K_m)_{m\geq l}$ in place of $(K_m)_{m\geq 1}$, to obtain a relatively closed subset F_l of Ω satisfying $K_l^\circ \cap F_l = \emptyset$ and $E \subset \Omega\backslash F_l$, and also a non-negative superharmonic function s_l on $\Omega\backslash F_l$ which is positive on the component of K_l° which contains Q and which satisfies $s_l < 2^{-l}\epsilon$ on E. If we define $F = \cup_l F_l$ and $s = \sum_l s_l$ on $\Omega\backslash F$, then F is a relatively closed subset of Ω satisfying $E \subset \Omega\backslash F$, and s is a positive superharmonic function on $\Omega\backslash F$ satisfying $s < \epsilon$ on E, as required.

Proof of Theorem 4.4. If (c) holds, then we can apply Theorem 3.17 and Lemma 4.5 to deduce (b). Clearly (b) implies (a). If (a) holds, then Theorem 1.6 shows that $\Omega\backslash E$ and $\Omega\backslash E^\circ$ are thin at the same points of E. It remains to show that (a) also implies the long islands condition.

To do this, suppose that (a) holds but the long islands condition fails. Then there is a sequence (V_k) of components (not necessarily distinct) of E° and two sequences (X_k), (Y_k) of points such that $X_k, Y_k \in V_k$ for each k, such that (X_k) converges to a point P in $\partial^*\Omega$, and such that (Y_k) converges to a point Q in $\partial E \cap \Omega$. By using a Kelvin transformation centred at P, if necessary, we can suppose that $X_k \to \infty$. (The transformed pair (Ω', E') would still satisfy (a) but not the long islands condition.) We can also assume that Q is the origin O and that $|Y_k| < k^{-1}$ for each k. As in the proof of Lemma 4.5 we may assume that Ω is Greenian. Next, let (B_k) be a sequence of pairwise disjoint closed balls in $\Omega\backslash E$ with centres Z_k such that $Z_k \to O$, let u_k be the capacitary potential on Ω valued 1 on B_k, and define

$$u(X) = \sum_{k=1}^{\infty} 2^{-k} u_k(X) \qquad (X \in \Omega).$$

By uniform convergence $u \in C(E) \cap \mathcal{H}(E^\circ)$. For each k, let ω_k be the unbounded connected open set defined by

$$\omega_k = \left(\bigcup_{m=k}^{\infty} V_m \right) \cup \{X : 0 < |X| < k^{-1} \text{ and } X \notin \cup_m B_m\},$$

and let ϵ_{ω_k} be as in Theorem 4.1. Further, let $\epsilon : [0, +\infty) \to (0, 1]$ be a continuous function satisfying

$$\epsilon(t) \leq \min\{\epsilon_{\omega_1}(t), \ldots, \epsilon_{\omega_k}(t)\} \qquad (t \in [k-1, k); k \in \mathbf{N}).$$

It follows from condition (a) that there is a harmonic function v on an open set W which contains E, such that $|v-u|<\epsilon$ on E. Since $O \in E$, there exists k' in $\mathbf{N}$ such that $B(O,1/k') \subset W$. We define

$$a = \sup\left\{\frac{|v(X)-u(X)|}{\epsilon_{\omega_{k'}}(|X|)} : X \in \overline{\omega_{k'}} \text{ and } |X| \le k'\right\}.$$

Observing that $\epsilon(t) \le \epsilon_{\omega_{k'}}(t)$ whenever $t \ge k'$, we obtain

$$\frac{|v(X)-u(X)|}{a+1} < \epsilon_{\omega_{k'}}(|X|) \qquad (X \in \omega_{k'}).$$

Since $v-u \in \mathcal{H}(\omega_{k'})$, we conclude from Theorem 4.1 that $v \equiv u$ in $\omega_{k'}$. This contradicts the mean value property of v on a neighbourhood of any B_k contained in $B(O,1/k')$. Hence the long islands condition must hold.

4.4 Carleman Approximation

The harmonic analogue of Nersesyan's Theorem will now be deduced.

Theorem 4.6. *The following are equivalent:*
(a) for each u in $C(E)\cap\mathcal{H}(E^\circ)$ and each continuous function $\epsilon : E \to (0,1]$, there exists v in $\mathcal{H}(\Omega)$ such that $|v-u|<\epsilon$ on E;
(b) for each u in $C(E)\cap\mathcal{H}(E^\circ)$ and each continuous function $\epsilon : E \to (0,1]$, there exists v in $\mathcal{H}(\Omega)$ such that $0<v-u<\epsilon$ on E;
(c) $\Omega\backslash\widehat{E}$ and $\Omega\backslash E^\circ$ are thin at the same points of E, and (Ω, E) satisfies both the (K,L)-condition and the long islands condition.

Proof. If (c) holds, then (b) follows from Theorem 3.19 and Lemma 4.5. Clearly (b) implies (a). If (a) holds, then (c) follows from Theorems 3.19 and 4.4.

4.5 Approximation of Functions in $\mathcal{H}(E)$

Comparing Theorem 3.17 with Theorem 4.4, and Theorem 3.19 with Theorem 4.6, we observe a similar pattern: the pairs (Ω, E) which permit approximation with an arbitrary error function are precisely those which both permit uniform approximation and satisfy the long islands condition. This pattern breaks down (at least when $n \ge 3$) when we consider an analogue of Theorem 3.15 for approximation with an arbitrary error function. We need to introduce a fine topology analogue of the long islands condition as follows. A pair (Ω, E) is said to satisfy the *fine long islands condition* if, for each compact subset K of Ω, there is a compact subset L of Ω which contains every fine component of the fine interior of E that intersects K. In the next example we make use of the observations that, if W' is a finely open set in $\mathbf{R}^{n-1} (n \geq 3)$, then $W' \times \mathbf{R}$ is finely open in $\mathbf{R}^n$, and each fine component of the finely open set $W' \times (a,b)$ is of the form $A' \times (a,b)$, where $A' \subseteq \mathbf{R}^{n-1}$. (The first assertion follows from Theorem 0.A and the fact that, if $u \in \mathcal{S}(\mathbf{R}^{n-1})$ and we define $v(X', x_n) = u(X')$, then $v \in \mathcal{S}(\mathbf{R}^n)$. The second assertion follows from the translational symmetry of $W' \times \mathbf{R}$.)

Example 4.7. Let $n \geq 3$, let $A' = [0,1]^{n-2} \times ([-1,1] \backslash \{0\})$, let $\{Y'_k : k \in \mathbf{N}\}$ be a dense subset of A', and define

$$u(X') = \sum_{k=1}^{\infty} 2^{-k} \phi_{n-1}(|X' - Y'_k|) \qquad (X' \in \mathbf{R}^{n-1})$$

and

$$F' = ([0,1]^{n-2} \times \{0\}) \cup \left\{ (X'', x_{n-1}) \in A' : u(X'', x_{n-1}) \leq \phi_{n-1}(|x_{n-1}|) \right\}.$$

Then F' is closed, and has non-empty interior with respect to the fine topology on $\mathbf{R}^{n-1}$ (see Example 1.2). It follows that $F' \times \mathbf{R}$ has non-empty fine interior in $\mathbf{R}^n$. We define

$$F_m = \{(2^{-m-1}X' + Q'_m, x_n) : X' \in F' \text{ and } 0 \leq x_n \leq m\} \qquad (m \in \mathbf{N}),$$

where $Q'_m = (2^{-m}, 0, \ldots, 0) \in \mathbf{R}^{n-1}$, and further define

$$E = (\{0\}^{n-1} \times [0, +\infty)) \cup \left(\bigcup_{m \in \mathbf{N}} F_m \right).$$

Then E is closed, but $(\mathbf{R}^n, E)$ does not satisfy the fine long islands condition, as can be seen by letting $K = [0,1]^n$. Of course, $(\mathbf{R}^n, E)$ satisfies the ordinary long islands condition, since $E^\circ = \phi$.

Theorem 4.8. *The following are equivalent:*
(a) for each u in $\mathcal{H}(E)$ and each continuous function $\epsilon : E \to (0,1]$, there exists v in $H(\Omega)$ such that $0 < v - u < \epsilon$ on E;
(b) $\Omega \backslash \widehat{E}$ and $\Omega \backslash E$ are thin at the same points of E, and (Ω, E) satisfies both the (K, L)-condition and the fine long islands condition.

The proof of Theorem 4.8 requires a little material from fine potential theory, for which we give specific references. We begin by presenting the following analogue of Lemma 4.5.

Lemma 4.9. *If* (Ω, E) *satisfies the fine long islands condition and* $\epsilon : E \to (0,1]$ *is continuous, then there exists* s *in* $\mathcal{S}^+(E)$ *such that* $s < \epsilon$ *on* E.

Proof of Lemma. Using the fine long islands condition we can construct an exhaustion (K_m) of Ω such that $K_1^\circ \neq \emptyset$ and such that, for each m, every fine component V of the fine interior of E which satisfies $\overline{V} \cap K_m \neq \emptyset$ also satisfies $\overline{V} \subset K_{m+1}^\circ$.

Let $A[l,m], F(m;k), \delta_m, Q$ and g' be as defined in the proof of Lemma 4.5. We observe that

$$R_{g'}^{F(m;k)}(X) \uparrow R_{g'}^{A[m,m+5]\backslash E}(X) \qquad (k \to \infty; X \in \Omega; m \in \mathbf{N}).$$

If $u \in \mathcal{S}^+(\Omega)$ and $u(X) \geq g'(X)$ on $A[m, m+5]\backslash E$, then the same inequality holds for points X of the set

$$\{X \in E : A[m, m+5]\backslash E \text{ is not thin at } X\}.$$

The construction of (K_m) and the fine minimum principle (see [Doo, 1.XI.19]) together show that $u \geq g'$ on $A[m+1, m+4]$, and hence

$$R_{g'}^{F(m;k)}(X) \uparrow g'(X) \qquad (k \to \infty; X \in E \cap A[m+1, m+4]).$$

Dini's theorem implies that this convergence is uniform on the given set, so there exists k_m in $\mathbf{N}$ such that

$$g'(X) \geq R_{g'}^{F(m;k_m)}(X) > g'(X) - \delta_{m+2} \qquad (X \in E \cap A[m+2, m+3]).$$

The remainder of the proof now proceeds exactly as the part of the proof of Lemma 4.5 which follows (4.1).

Proof of Theorem 4.8. If (b) holds, then we can apply Theorem 3.15 and Lemma 4.9 to deduce (a). Conversely, if (a) holds, then it follows from Theorem 3.15 that $\Omega\backslash\widehat{E}$ and $\Omega\backslash E$ are thin at the same points of E, and that (Ω, E) satisfies the (K, L)-condition. It remains to show that (a) also implies the fine long islands condition.

To do this, suppose that (a) holds, but that the fine long islands condition fails. Then there is a compact subset K of Ω, a sequence (V_k) of fine components (not necessarily distinct) of the fine interior of E which satisfy $V_k \cap K \neq \emptyset$, and a sequence (X_k) of points such that $X_k \in V_k$ for each k and $X_k \to \mathcal{A}$. Let U be an Ω-bounded connected open set which contains K, and define $\omega = U \cup (\cup_k V_k)$. Then ω is a finely connected finely open set.

Let s be the fine balayage of the constant function 1 relative to U in the finely open set ω (see [Fug1, §11]), and let $\delta_k = 2^{-k}s(X_k)$. Since ω is finely connected, we know (see [Fug1, Theorem 12.6]) that $\delta_k > 0$ for each k, and so we can choose a continuous function $\epsilon : E \to (0,1]$ such that $\epsilon(X_k) \leq \delta_k$ for each k.

If we define $u \equiv 0$, then by (a) there exists v in $\mathcal{H}(\Omega)$ such that $0 < v < \epsilon$ on E. Let $a = \inf\{v(X) : X \in \overline{U} \cap E\}$ and

$$w(X) = \begin{cases} \min\{v(X), a/2\} & (X \in \Omega \backslash U) \\ a/2 & (X \in U). \end{cases}$$

Then $a > 0$ and $w \in \mathcal{S}^+(\omega)$. Clearly $w \geq (a/2)s$ on ω, so

$$2^{-k}s(X_k) = \delta_k \geq \epsilon(X_k) > v(X_k) \geq w(X_k) \geq (a/2)s(X_k) \qquad (k \in \mathbf{N}),$$

which yields a contradiction. Thus the fine long islands condition must hold.

As usual, the two-dimensional case is simpler to describe.

Theorem 4.10. *Let $n = 2$. The following are equivalent:*
(a) for each u in $\mathcal{H}(E)$ and each continuous function $\epsilon : E \to (0,1]$, there exists v in $\mathcal{H}(\Omega)$ such that $|v - u| < \epsilon$ on E;
(b) for each u in $C(E) \cap \mathcal{H}(E^\circ)$ and each continuous function $\epsilon : E \to (0,1]$, there exists v in $\mathcal{H}(\Omega)$ such that $0 < v - u < \epsilon$ on E;
(c) $\partial \widehat{E} = \partial E$ and (Ω, E) satisfies both the (K, L)-condition and the long islands condition.

Proof. If (c) holds, then (b) follows from Corollary 3.21, Lemma 4.5 and the argument used to prove Corollary 1.16. Clearly (b) implies (a). If (a) holds, then Corollary 3.21 shows that $\partial \widehat{E} = \partial E$ and that (Ω, E) satisfies the (K, L)-condition. Finally, it follows from Theorem 4.8 that (Ω, E) satisfies the fine long islands condition, and hence the long islands condition (see [Fug1, p. 87]).

Problem 4.11. Let $n \geq 3$. Characterize those pairs (Ω, E) which have the following property: for each u in $\mathcal{H}(E)$ and each continuous function $\epsilon : E \to (0,1]$, there exists v in $\mathcal{H}(\Omega)$ such that $|v - u| < \epsilon$ on E.

Notes

An early result concerning Carleman-type harmonic approximation, which dealt with the special case of sets E with empty interior, can be found in

[Sha1]. As we noted in §4.5, a comparison of Theorem 3.19 with Theorem 4.6 shows that the pairs (Ω, E) which permit approximation with an arbitrary error function are precisely those which permit uniform approximation and which also satisfy the long islands condition. This fact was first established in the case where $n = 2$ by Bagby and Gauthier [BG1]. The corresponding result in higher dimensions, which cannot be proved by the methods of [BG1], was first established by Gardiner and Goldstein [GG] using an argument different from that presented here. (The necessity of the long islands condition in any dimension was independently shown in [BG3].) The above exposition is based on [Gar5]. Theorem 4.1 is due to Armitage, Bagby and Gauthier [ABG]. In connection with Lemma 4.2 we mention that a more precise "transfer of smallness" result has recently been obtained by Korevaar and Meyers [KM]. Theorems 4.4, 4.6, 4.8 can be found in [GG], [Gar5]. The equivalence of (b) and (c) in Theorem 4.4 has been established in the context of harmonic spaces (where fusion techniques are not available) by Gardiner, Goldstein and GowriSankaran [GGG2]. In connection with Problem 4.11 we mention that the fine topology analogue of Theorem 4.1 does not hold: Lyons [Lyo] has shown that finely harmonic functions need not be quasi-analytic.

5 Tangential Approximation at Infinity

5.1 Introduction

If E is a closed subset of $\mathbf{R}^n$ such that $(\mathbf{R}^n)^* \backslash E$ is connected and locally connected, then it follows from Corollary 3.10 that for every u in $\mathcal{H}(E)$ and every positive number ϵ, there exists v in $\mathcal{H}(\mathbf{R}^n)$ such that

$$|v(X) - u(X)| < \epsilon\big(1 + |X|\big)^{2-n} \qquad (X \in E).$$

However, for some applications (see §8.1), the above decay in the error of approximation is not sufficiently rapid. In this chapter it will be shown that, for any positive number a, it is possible to approximate on E with an error of at most $\epsilon\big(1+|X|\big)^{-a}$. The results below complement, rather than improve, those of Chapter 3. Here we are arranging for the error of approximation to decay at a certain rate at ∞, whereas in Chapter 3 we worked in the context of a Greenian open set Ω and arranged for the error to vanish at all regular points of $\partial^*\Omega$.

5.2 Preliminary Lemmas

The first main aim of this chapter is to establish a modified form of the fusion result in Chapter 2. For this we require some preliminary lemmas and definitions.

Lemma 5.1. *Let $a \in \mathbf{N}$. There exists a positive constant $c(n, a)$ such that, if $u \in \mathcal{H}(\mathbf{R}^n \backslash B(O, r))$ and $|X|^a u(X)$ is bounded on $\mathbf{R}^n \backslash B(O, r)$, then*

$$|u(X)| \le c(n,a)\big(r/|X|\big)^a \sup\Big\{|u(X)| : X \in \partial B(O,r)\Big\} \qquad \big(|X| \ge 2r\big).$$

Proof. We present the proof in the case where $n \ge 3$; a routine modification is required when $n = 2$. We may assume that $r = 1$. The Laurent expansion of u about O can be written as

$$u(X) = h(X) + \sum_{k=0}^{\infty} |X|^{2-n-2k} H_k(X) \qquad (X \in \mathbf{R}^n \backslash B(O,1)),$$

where $h \in \mathcal{H}(\mathbf{R}^n)$ and $H_k \in \mathcal{H}_k$ for each k. Let

$$v_l(X) = \sum_{k=l}^{\infty} |X|^{2-n-2k} H_k(X) \qquad (l = 0, 1, \ldots).$$

It follows (see the proof of Lemma 2.5) from the homogeneity of each H_k that the function $|X|^{n-2+l} v_l(X)$ is bounded on $\mathbf{R}^n \backslash B(O,1)$. Since the function $|X|^a u(X)$ is bounded, we can conclude that $h(X) \to 0$ as $|X| \to \infty$, and so $h \equiv 0$. Similarly $H_k \equiv 0$ when $k \in \{0, 1, \ldots, l'-1\}$, where l' is given by $l' = \max\{a + 2 - n, 0\}$. Hence $u = v_{l'}$. Denoting the mean value of a function f over $\partial B(O,\rho)$ by $\mathcal{M}(f; O, \rho)$, we recall that $\mathcal{M}(H_j H_k; O, \rho) = 0$ whenever $j \neq k$ (see, for example, [DuP, Lemma 2.43]). Hence

$$\mathcal{M}(u^2; O, \rho) = \sum_{k=l'}^{\infty} \rho^{4-2n-4k} \mathcal{M}(H_k^2; O, \rho) = \sum_{k=l'}^{\infty} b_k \rho^{4-2n-2k} \qquad (\rho \geq 1),$$

where $b_k = \mathcal{M}(H_k^2; O, 1)$, and we deduce that

$$\begin{aligned} \mathcal{M}(u^2; O, \rho) &\leq \rho^{-2a} \sum_{k=l'}^{\infty} b_k \\ &= \rho^{-2a} \mathcal{M}(u^2; O, 1) \\ &\leq \rho^{-2a} \Big(\sup\{|u(X)| : X \in \partial B(O,1)\} \Big)^2 \end{aligned}$$

when $\rho \geq 1$. Hence, if $|X| \geq 2$ and σ_n denotes $\sigma\big(\partial B(O,1)\big)$ as usual, we can use the subharmonicity of u^2 to obtain

$$\begin{aligned} \{u(X)\}^2 &\leq \frac{2^n n}{\sigma_n |X|^n} \int_{B(X,|X|/2)} \{u(Y)\}^2 d\lambda(Y) \\ &\leq \frac{2^n n}{|X|^n} \int_{|X|/2}^{3|X|/2} t^{n-1} \mathcal{M}(u^2; O, t)\, dt \\ &\leq c(n,a) |X|^{-2a} \Big(\sup\{|u(X)| : X \in \partial B(O,1)\} \Big)^2, \end{aligned}$$

from which the lemma follows.

We next describe a modification of the fundamental function $\phi_n(|X - Y|)$ which is essentially the same as that given in Hayman and Kennedy [HK,§4.1]. For each k in $\mathbf{N}$, let $L_{k,n}$ be the generalized Legendre polynomial of degree k which arises as the solution of the differential equation

$$\left\{(1-t^2)\frac{d^2}{dt^2}-(n-1)t\frac{d}{dt}+k(k+n-2)\right\}L_{k,n}(t)=0 \qquad (t\in\mathbf{R})$$

subject to the conditions $L_{k,n}(1)=1$ and $L_{k,n}(-1)=(-1)^k$. A convenient account of these polynomials can be found in the lecture notes by Müller [Mul]. We note that $L_{k,n}$ is proportional to the Jacobi polynomial $P_k^{(\alpha,\alpha)}$ where $\alpha=(n-3)/2$, and also (when $n\geq 3$) to the ultraspherical polynomial $P_k^{(\lambda)}$ where $\lambda=(n-2)/2$ (see [Sze, Chapter IV] and [DuP, §1.6]). The particular facts that we shall require are summarized below.

Lemma 5.A.
(i) If $X\in\mathbf{R}^n\backslash\{O\}$ *, then the function*

$$Y\mapsto |Y|^k L_{k,n}(\langle X,Y\rangle/(|X|\,|Y|)) \qquad (Y\in\mathbf{R}^n)$$

belongs to $\mathcal{H}_k$.
(ii) If $t\in[-1,1]$*, then* $|L_{k,n}(t)|\leq 1$.
(iii) If $x\in[0,1)$ *and* $t\in[-1,1]$*, then*

$$\phi_n\big((1+x^2-2xt)^{1/2}\big)-\phi_n(1)=\sum_{k=1}^{\infty}c_{k,n}x^kL_{k,n}(t),$$

where $c_{0,2}=1$, $c_{k,2}=-k^{-1}$ *when* $k\geq 1$, *and*

$$c_{k,n}=\binom{k+n-3}{k} \qquad (n\geq 3, k\geq 0).$$

We refer to [Mul, Theorem 1 and Lemma 23] for (i) above, to [Mul, Lemma 9] for (ii), and to [Mul, Lemma 18] for (iii).

We now define functions on $\mathbf{R}^n\times\mathbf{R}^n$ by

$$P_0(X,Y)=\begin{cases}\phi_n(|X|) & (X\neq O)\\ 0 & (X=O)\end{cases}$$

and, if $k\in\mathbf{N}$, by

$$P_k(X,Y)=\begin{cases}|Y|^k|X|^{2-n-k}L_{k,n}\left(\dfrac{\langle X,Y\rangle}{(|X|\,|Y|)}\right) & (X\neq O \text{ and } Y\neq O)\\ 0 & (X=O \text{ or } Y=O)\end{cases}$$

It follows from Lemma 5.A(i) that, if $k\geq 1$, then $P_k(X,\,\cdot\,)\in\mathcal{H}_k$ for each X in $\mathbf{R}^n$, and hence the function

$$X\mapsto |X|^{2k+n-2}P_k(X,Y)$$

belongs to $\mathcal{H}_k$ for each Y in $\mathbf{R}^n$. It follows easily from the Kelvin transformation that $P_k(\,\cdot\,, Y) \in \mathcal{H}(\mathbf{R}^n\backslash\{O\})$ for each Y in $\mathbf{R}^n$. Also, clearly $P_k \in C^\infty\big((\mathbf{R}^n\backslash\{O\}) \times \mathbf{R}^n\big)$ when $k \geq 0$, since

$$(X,Y) \mapsto |X|^{2k+n-2} P_k(X,Y)$$

defines a polynomial on $\mathbf{R}^n \times \mathbf{R}^n$. If $X, Y \in \mathbf{R}^n$ and if we write $x = |Y| \,/\, |X|$ and $t = \langle X, Y\rangle \,/\, \big(|X|\,|Y|\big)$, then we see from Lemma 5.A(iii) that

$$\sum_{k=0}^{\infty} c_{k,n} P_k(X,Y) = \phi_n\big(|X-Y|\big) \qquad \big(|X| > |Y| > 0\big). \tag{5.1}$$

Next, for each l in $\mathbf{N} \cup \{0\}$, we define

$$Q_l(X,Y) = \sum_{k=0}^{l} c_{k,n} P_k(X,Y)$$

and

$$R_l(X,Y) = \begin{cases} \phi_n\big(|X-Y|\big) - Q_l(X,Y) & (Y \neq O) \\ 0 & (Y = O). \end{cases}$$

In view of what was said above, we see that $Q_l \in C^\infty\big((\mathbf{R}^n\backslash\{O\}) \times \mathbf{R}^n\big)$, that $Q_l(X,\,\cdot\,) \in \mathcal{H}(\mathbf{R}^n)$ and that $D_2^\alpha Q_l(\,\cdot\,, Y) \in \mathcal{H}(\mathbf{R}^n\backslash\{O\})$, where $\alpha = (\alpha_1, \alpha_2, \ldots, \alpha_n)$ and

$$D_2^\alpha = \frac{\partial^{\alpha_1 + \cdots + \alpha_n}}{\partial y_1^{\alpha_1} \ldots \partial y_n^{\alpha_n}}.$$

Lemma 5.2. *If* $l \in \mathbf{N} \cup \{0\}$, *then*

$$\big|R_l(X,Y)\big| \leq c(n,l)|Y|^{l+1}|X|^{1-n-l} \qquad \big(0 \leq |Y| < |X|/2\big) \tag{5.2}$$

and

$$\big|\nabla_2 R_l(X,Y)\big| \leq c(n,l)|Y|^{l}|X|^{1-n-l} \qquad \big(0 \leq |Y| < |X|/4\big). \tag{5.3}$$

Proof. If $0 < |Y| < |X|/2$, then it follows from (5.1), the definition of $P_k(\,\cdot\,,\,\cdot\,)$ and Lemma 5.A(ii) that

$$\begin{aligned} |X|^{n+l-1}\big|R_l(X,Y)\big| &\leq |X|^{n+l-1} \sum_{k=l+1}^{\infty} |c_{k,n}|\,\big|P_k(X,Y)\big| \\ &\leq \sum_{k=l+1}^{\infty} |c_{k,n}|\,|Y|^k |X|^{l-k+1} \\ &\leq |Y|^{l+1} \sum_{k=l+1}^{\infty} |c_{k,n}| 2^{l-k+1}, \end{aligned}$$

whence (5.2) holds. (The inequality is trivial when $Y = O$.) Inequality (5.3) follows from (5.2) and the fact that $R_l(X,\cdot) \in \mathcal{H}(\mathbf{R}^n \backslash \{X\})$, using Harnack's inequality (cf. (2.1)).

Now let U be a bounded admissible open set, let $l \in \mathbf{N} \cup \{0\}$, let $\psi \in C^1(\overline{U}) \cap C^2(U)$ and $\Omega_0 = \mathbf{R}^n \setminus \big((\partial U \cap \operatorname{supp} \psi) \cup \{O\}\big)$. As in Chapter 2 we obtain a harmonic function on Ω_0 by defining

$$w(X) = c_n \int_{\partial U} \left\{ \psi(Y) \frac{\partial}{\partial n_Y} R_l(X,Y) - R_l(X,Y) \frac{\partial \psi}{\partial n_Y}(Y) \right\} d\sigma(Y) \quad (X \in \Omega_0).$$

Lemma 5.3. *Let U, l, ψ and w be as above, let ω be a bounded open set which contains $\partial U \cap \operatorname{supp} \psi$, let $\epsilon > 0$, and let $\rho > 0$. Then there exists w_1 in $\mathcal{I}(\mathbf{R}^n) \cap \mathcal{H}(\Omega_0)$ such that*

$$|w_1(X) - w(X)| < \epsilon\big(1 + |X|\big)^{1-n-l} \qquad \Big(X \in \mathbf{R}^n \backslash \big(\omega \cup B(O,\rho)\big)\Big).$$

Proof. Let r be large enough so that $\overline{\omega} \subset B(O, r/4)$ and $r > \rho$. We approximate the integral defining w above by a Riemann sum w_1 in such a way that

$$|w_1(X) - w(X)| < (\epsilon/C)(1 + 2r)^{1-n-l} \qquad (X \in \mathbf{R}^n \backslash \omega; \rho \le |X| \le 2r), \tag{5.4}$$

where C is the positive constant of Lemma 5.1 corresponding to $a = n+l-1$. (We may assume that $C \ge 1$). It follows from Lemma 5.2 that

$$|w(X)| \le c(\partial U \cap \operatorname{supp} \psi, n, l) c(\psi) |X|^{1-n-l} \qquad (|X| \ge r),$$

and a similar inequality holds for w_1. Hence, by Lemma 5.1,

$$\begin{aligned} |w_1(X) - w(X)| &\le \epsilon\big(|X|(1+2r)/r\big)^{1-n-l} \\ &\le \epsilon(1 + |X|)^{1-n-l} \qquad (|X| \ge 2r). \end{aligned}$$

Combining the above inequality with (5.4), we obtain the conclusion of the lemma.

Lemma 5.4. *Let U, l, ψ and w be as above. Then*

$$w(X) + c_n \int_U R_l(X,Y) \Delta\psi(Y) d\lambda(Y) = \begin{cases} -\psi(X) & (X \in U) \\ 0 & (X \in (\mathbf{R}^n \backslash \overline{U}) \cup (\partial U \backslash \operatorname{supp} \psi)). \end{cases}$$

Proof. Using the fact that $Q_l(X,\cdot) \in \mathcal{H}(\mathbf{R}^n)$ for each X, we simply repeat the argument given for Lemma 2.4.

5.3 A Fusion Result

Theorem 5.5 *Let $a > 0$, let K, E_1 be compact sets and let E_2 be a closed set such that $E_1 \cap E_2 = \emptyset$ and $K \cup E_1 \cup E_2 \neq \mathbf{R}^n$. Then there is a positive constant C with the following property: if $u_1, u_2 \in \mathcal{I}(\mathbf{R}^n) \cap \mathcal{H}(K)$ and $|u_1 - u_2| < \epsilon$ on K, then there exists u in $\mathcal{I}(\mathbf{R}^n)$ such that*

$$\left|(u-u_k)(X)\right| < C\epsilon\left(1+|X|\right)^{-a} \qquad (X \in K \cup E_k; k = 1,2). \tag{5.5}$$

Proof. As in §2.3, we may assume that $u_2 \equiv 0$ and $K \neq \emptyset$. We may also assume, without loss of generality, that there exists a positive number ρ such that $\overline{B(O,\rho)}$ does not intersect $K \cup E_1 \cup E_2$. Let ω_1, ω_2 be bounded admissible open sets such that $E_1 \subset \omega_1$, $\overline{\omega}_1 \subset \omega_2$ and $\overline{\omega}_2 \subset \mathbf{R}^n \backslash E_2$. Also, let r be a positive number such that $K \subset B(O,r)$. Since $u_1 \in \mathcal{H}(K)$ and $|u_1| < \epsilon$ on K, there is an open set V such that $K \subset V \subseteq B(O,r)\backslash\overline{B(O,\rho)}$ and $\sup_V |u_1| < \epsilon$. Further, it can be arranged that $\omega_1 \cup V$ is admissible, and that $u_1 \in \mathcal{H}(\overline{V})$. Also, we may assume that u_1 has no singularities on the set $\partial\omega_1 \cup \partial\omega_2$ (see the second paragraph of the proof of Theorem 2.7, and use Lemma 2.5 in place of Lemma 2.6). We choose ϕ in $C^\infty(\mathbf{R}^n)$ such that $0 \leq \phi \leq 1$ on $\mathbf{R}^n$, $\phi = 1$ on ω_1 and $\phi = 0$ on $\mathbf{R}^n\backslash\omega_2$, and define

$$\psi = \begin{cases} \phi u_1 & \text{on } \omega_2 \\ 0 & \text{on } \mathbf{R}^n\backslash\omega_2. \end{cases}$$

Thus $\psi = u_1$ on E_1, $\psi = 0$ on E_2 and $|\psi| \leq |u_1|$ on K. Since $|u_1| < \epsilon$ on $\overline{V}$, we obtain

$$\left|(\psi - u_1)(X)\right| \leq C_1\epsilon(1+|X|)^{-a} \qquad (X \in K \cup E_1) \tag{5.6}$$

and

$$\left|\psi(X)\right| \leq C_1\epsilon\left(1+|X|\right)^{-a} \qquad (X \in K \cup E_2), \tag{5.7}$$

where $C_1 = (1+r)^a$.

Let S denote the (finite) set of singularities of u_1 in ω_1, and define $S_\delta = \{X : \operatorname{dist}(X,S) \leq \delta\}$, where δ is small enough to ensure that S_δ is the disjoint union of closed balls contained in ω_1. Let $W = \omega_1 \cup V \cup (\mathbf{R}^n\backslash\overline{\omega_2})$ and choose l in $\mathbf{N}$ such that $l > a+1-n$. We now follow the proof of Theorem 2.7 (with Lemma 5.4 in place of Lemma 2.4, and $R_l(\cdot,\cdot)$ in place of $G(\cdot,\cdot)$) to obtain

$$\psi(X) = s_2(X) + s_3(X) + s_4(X)$$
$$+ c_n \int_V u_1(Y)\Big\{R_l(X,Y)\Delta\phi(Y) + 2\big\langle \nabla_2 R_l(X,Y), \nabla\phi(Y)\big\rangle\Big\} d\lambda(Y)$$
$$(X \in W\backslash S),$$

where

$$s_2(X) = -c_n \int_{\partial W} \left\{\psi(Y)\frac{\partial}{\partial n_Y} R_l(X,Y) - R_l(X,Y)\frac{\partial\psi}{\partial n_Y}(Y)\right\} d\sigma(Y),$$
$$s_3(X) = c_n \int_{\partial S_\delta} \left\{\psi(Y)\frac{\partial}{\partial n_Y} R_l(X,Y) - R_l(X,Y)\frac{\partial\psi}{\partial n_Y}(Y)\right\} d\sigma(Y)$$

$\big(s_3 \in \mathcal{H}(\mathbf{R}^n\backslash S)\big)$, and

$$s_4(X) = -2c_n \int_{\partial W} u_1(Y) R_l(X,Y) \frac{\partial\phi}{\partial n_Y}(Y)\, d\sigma(Y).$$

Hence

$$\big|(s_2 + s_3 + s_4)(X) - \psi(X)\big| \le C_2 \epsilon F(X) \qquad (X \in W\backslash S), \tag{5.8}$$

where C_2 is a positive constant independent of u_1, and

$$F(X) = \int_{B(O,r)\backslash \overline{B(O,\rho)}} \Big\{\big|R_l(X,Y)\big| + \big|\nabla_2 R_l(X,Y)\big|\Big\} d\lambda(Y).$$

If $|X| > 4r$, then we can use Lemma 5.2 to see that $\big|F(X)\big| \le C_3|X|^{-a}$, where C_3 depends on n, l and r. Recalling that $Q_l \in C^\infty\Big(\big(\mathbf{R}^n\backslash\{O\}\big)\times\mathbf{R}^n\Big)$, we obtain

$$F(X) \le C_4 + C_5 \int_{B(O,r)} \{\phi_n(|X-Y|) + |X-Y|^{1-n}\}\, d\lambda(Y)$$
$$\le C_4 + C_5 \int_{B(X,r)} \Big\{\phi_n\big(|X-Y|\big) + |X-Y|^{1-n}\Big\}\, d\lambda(Y)$$
$$\le C_6 \qquad (\rho \le |X| \le 4r),$$

where C_4, C_5 and C_6 depend at most on n, l, ρ and r. Hence

$$F(X) \le C_7\big(1 + |X|\big)^{-a} \qquad \big(|X| \ge \rho\big),$$

and it follows from (5.8) that

$$\big|(s_2 + s_3 + s_4)(X) - \psi(X)\big| \le C_8\epsilon(1 + |X|)^{-a} \qquad (X \in W\backslash S).$$

We noted above that $s_3 \in \mathcal{H}(\mathbf{R}^n \backslash S)$. Also, there exist w_2 and w_4 in $\mathcal{I}(\mathbf{R}^n)$ such that

$$\left|(w_k - s_k)(X)\right| < \epsilon\big(1 + |X|\big)^{-a} \qquad (X \in K \cup E_1 \cup E_2; k = 2, 4):$$

this is a consequence of the reasoning used to prove Lemma 5.3. If we now define $u = w_2 + s_3 + w_4$, then $u \in \mathcal{I}(\mathbf{R}^n)$ and we see that

$$\left|(u - \psi)(X)\right| < (C_8 + 2)\epsilon\big(1 + |X|\big)^{-a} \qquad (X \in (K \cup E_1 \cup E_2) \backslash S).$$

Combining this with (5.6) and (5.7) we obtain (5.5). The theorem is now proved.

5.4 Pole Pushing

The results of this section are in preparation for the approximation theorems of §§5.5, 5.6. Recall that, if $u \in \mathcal{I}(\mathbf{R}^n)$ and Y is a singularity of u, then an open set T is called a *tract for* u *and* Y if $u \in \mathcal{H}\big(T \backslash \{Y\}\big)$ and T contains a path connecting Y to ∞.

Lemma 5.6. *Suppose that* $u \in \mathcal{I}(\mathbf{R}^n)$, *that* $a > 0$ *and* $\epsilon > 0$, *and that* Y_0 *is a singularity of* u. *If* T *is a tract for* u *and* Y_0, *then there exists* v *in* $\mathcal{I}(\mathbf{R}^n) \cap \mathcal{H}(T)$ *such that*

$$\left|(v - u)(X)\right| < \epsilon\big(1 + |X|\big)^{-a} \qquad (X \in \mathbf{R}^n \backslash T).$$

Proof. This follows from repeated application of Lemma 2.5 just as Lemma 3.4 was obtained by repeated application of Lemma 2.6.

Theorem 5.7. *Let* E *be a closed set such that* $(\mathbf{R}^n)^* \backslash E$ *is connected and locally connected. Then, for each* u *in* $\mathcal{I}(\mathbf{R}^n) \cap \mathcal{H}(E)$ *and each choice of positive numbers* a *and* ϵ, *there exists* v *in* $\mathcal{H}(\mathbf{R}^n)$ *such that*

$$\left|(v - u)(X)\right| < \epsilon\big(1 + |X|\big)^{-a} \qquad (X \in E).$$

Proof. This is proved using Lemma 5.6 just as Theorem 3.3 was proved using Lemma 3.4.

Lemma 5.8. *If* $u \in \mathcal{I}\big(\mathbf{R}^n \backslash \{O\}\big)$ *and* ϵ, ρ, a *are positive numbers, then there exists* v *in* $\mathcal{I}(\mathbf{R}^n)$ *such that*

$$\left|(v-u)(X)\right| < \epsilon\left(1+|X|\right)^{-a} \qquad \left(|X|>\rho\right). \tag{5.9}$$

Proof. Let S denote the set of singularities of u in $B(O,\rho/2)\backslash\{O\}$. It follows from Corollary 3.9 that there exists w_1 in $\mathcal{H}(\mathbf{R}^n\backslash(S\cup\{O\}))$ such that $u-w_1$ has only removable singularities at points of S. By Lemma 2.5 there exists w_2 in $\mathcal{H}\left(\mathbf{R}^n\backslash\{O\}\right)$ such that

$$\left|(w_2-w_1)(X)\right| < \epsilon\left(1+|X|\right)^{-a} \qquad \left(|X|>\rho\right).$$

If we define $v=(u-w_1)+w_2$, then $v\in\mathcal{I}(\mathbf{R}^n)$ and (5.9) holds.

5.5 Approximation of Functions in $\mathcal{H}(E)$

From now on E denotes a closed set in $\mathbf{R}^n$, and $\widehat{E}$ denotes the union of E with all bounded components of $\mathbf{R}^n\backslash E$.

Theorem 5.9. *The following are equivalent:*
(a) for each u in $\mathcal{H}(E)$ and each positive number ϵ, there exists v in $\mathcal{H}(\mathbf{R}^n)$ such that $|v-u|<\epsilon$ on E;
(b) for each u in $\mathcal{H}(E)$ and each choice of positive numbers ϵ and a, there exists v in $\mathcal{H}(\mathbf{R}^n)$ such that

$$\left|(v-u)(X)\right| < \epsilon\left(1+|X|\right)^{-a} \qquad (X\in E); \tag{5.10}$$

(c) $\mathbf{R}^n\backslash\widehat{E}$ and $\mathbf{R}^n\backslash E$ are thin at the same points of E, and $(\mathbf{R}^n,E)$ satisfies the (K,L)-condition.

Corollary 5.10. *If $(\mathbf{R}^n)^*\backslash E$ is connected and locally connected, then for each u in $\mathcal{H}(E)$ and each choice of positive numbers a and ϵ, there exists v in $\mathcal{H}(\mathbf{R}^n)$ such that (5.10) holds.*

The Corollary is immediate, since the hypotheses imply that condition (c) of Theorem 5.9 holds.

Proof of Theorem 5.9. Clearly (b) implies (a), and it follows from Theorem 3.15 that (a) implies (c). Now suppose that (c) holds, let $u\in\mathcal{H}(E)$, let $a>0$ and $\epsilon>0$. If $E=\mathbf{R}^n$, then (b) trivially holds. Thus we may suppose, without loss of generality, that $\overline{B(O,\rho)}\cap\widehat{E}=\emptyset$ for some positive number ρ. (If (c) holds and $\widehat{E}=\mathbf{R}^n$, then $E=\mathbf{R}^n$.)

Let (K_m) be an exhaustion of $\mathbf{R}^n\backslash\{O\}$, and let (ρ_m) be a sequence of positive numbers such that $K_m\subset B(O,\rho_m)$ for each m. Further, let

$F_m = K_m \cap \widehat{E}$, and let C_m be the positive constant of Theorem 5.5 corresponding to the assignments $K = F_{m+1}, E_1 = K_m$ and $E_2 = \widehat{E} \backslash K^\circ_{m+1}$. We may assume that $C_m \geq 1$, and we define (δ_m) to be a decreasing sequence of positive numbers such that $\delta_m \leq 2^{-m-2}\epsilon / C_m$. It follows from Theorem 3.15 that, for each m, there exists q_m in $\mathcal{H}(\mathbf{R}^n)$ such that

$$\left|(q_m - u)(X)\right| < \delta_m (1 + \rho_m)^{-a} \qquad (X \in E), \tag{5.11}$$

whence, by the maximum principle,

$$\left|(q_{m+1} - q_m)(X)\right| < 2\delta_m \qquad (X \in F_{m+1}).$$

Thus we can apply Theorem 5.5 to obtain r_m in $\mathcal{I}(\mathbf{R}^n)$ such that

$$\left|(r_m - q_m)(X)\right| < 2^{-m-1}\epsilon(1 + |X|)^{-a} \qquad (X \in K_m \cup F_{m+1}) \tag{5.12}$$

and

$$\left|(r_m - q_{m+1})(X)\right| < 2^{-m-1}\epsilon(1 + |X|)^{-a} \qquad (X \in \widehat{E}). \tag{5.13}$$

It follows that $r_m \in \mathcal{H}(\widehat{E})$, and that $r_m - q_m \in \mathcal{H}(K_m)$ for each m. Thus, if we define

$$v_1 = q_1 + \sum_{k=1}^{\infty} (r_k - q_k),$$

we see that $v_1 \in \mathcal{I}(\mathbf{R}^n \backslash \{O\}) \cap \mathcal{H}(\widehat{E})$. If $X \in E \cap K_m$, then

$$\begin{aligned} \left|(v_1 - u)(X)\right| &\leq \sum_{k=1}^{m-1} \left|(r_k - q_{k+1})(X)\right| + \left|(q_m - u)(X)\right| \\ &\quad + \sum_{k=m}^{\infty} \left|(r_k - q_k)(X)\right| \\ &< (5\epsilon/8)\big(1 + |X|\big)^{-a}, \end{aligned} \tag{5.14}$$

using (5.11)–(5.13). Since this holds for each m, (5.14) must hold for all X in E.

Now we apply Lemma 5.8 to obtain v_2 in $\mathcal{I}(\mathbf{R}^n)$ such that

$$\left|(v_2 - v_1)(X)\right| < (\epsilon/8)\big(1 + |X|\big)^{-a} \tag{5.15}$$

when $|X| > \rho$. In particular, (5.15) must hold for all X in $\widehat{E}$, and so $v_2 \in \mathcal{H}(\widehat{E})$. Finally, since $(\mathbf{R}^n)^* \backslash \widehat{E}$ is connected and (by (c)) locally connected, we can apply Theorem 5.7 to obtain v_3 in $\mathcal{H}(\mathbf{R}^n)$ such that

$$\left|(v_3 - v_2)(X)\right| < (\epsilon/4)\big(1 + |X|\big)^{-a} \qquad (X \in \widehat{E}). \tag{5.16}$$

Combining (5.14)–(5.16), we obtain (5.10). The theorem is now proved.

5.6 Approximation of Functions in $C(E) \cap \mathcal{H}(E^\circ)$

Theorem 5.11. *The following are equivalent:*
(a) for each u in $C(E) \cap \mathcal{H}(E^\circ)$ and each positive number ϵ there exists v in $\mathcal{H}(E)$ such that $|v - u| < \epsilon$ on E;
(b) for each u in $C(E) \cap \mathcal{H}(E^\circ)$ and each choice of positive numbers a and ϵ there exists v in $\mathcal{H}(E)$ such that

$$\left|(v-u)(X)\right| < \epsilon\left(1+|X|\right)^{-a} \qquad (X \in E); \tag{5.17}$$

(c) $\mathbf{R}^n \backslash E$ and $\mathbf{R}^n \backslash E^\circ$ are thin at the same points of E.

Proof. Clearly (b) implies (a), and Theorem 1.6 shows that (a) implies (c). Now suppose that (c) holds, let $a > 0$ and $\epsilon > 0$. We may suppose that $E \neq \mathbf{R}^n$ and $O \notin E$. Let (K_m) be an exhaustion of $\mathbf{R}^n \backslash \{O\}$ such that $\mathbf{R}^n \backslash K_m$ is not thin at any point of ∂K_m. We define $F_m = K_m \cap E$ for each m, and observe from condition (c) that $\mathbf{R}^n \backslash F_m$ and $\mathbf{R}^n \backslash F_m^\circ$ are thin at the same points. It follows from Theorem 1.3 and Lemma 1.8 that, for each positive number η and each natural number m there exists q in $\mathcal{I}(\mathbf{R}^n)$ such that

$$\left|(q-u)(X)\right| < \eta\left(1+|X|\right)^{-a} \qquad (X \in F_m).$$

With this extra ingredient we now follow the proof of Theorem 3.5 (using Theorem 5.5 in place of Theorem 2.7) to obtain v in $\mathcal{I}(\mathbf{R}^n) \cap \mathcal{H}(E)$ such that (5.17) holds.

The following example shows that the decay in the error bound in Theorem 5.11 cannot, in general, be improved.

Example 5.12. Let $\Omega = \mathbf{R}^n$ if $n \geq 3$, and let $\Omega = \mathbf{R}^2 \backslash \overline{B(O,1)}$ if $n = 2$. Further, let $E = \mathbf{R}^n \backslash B(O,2)$ and let $f : [0, +\infty) \to (0,1]$ be such that $t^a f(t) \to 0$ as $t \to +\infty$ for any choice of positive number a. For each k in $\mathbf{N}$ we define u_k to be the capacitary potential on Ω valued 1 on the closed ball B_k of centre $(0, \ldots, 0, 2 - 2^{-k})$ and radius 2^{-k-2}, and define

$$u(X) = \sum_{k=1}^{\infty} 2^{-k} u_k(X) \qquad (X \in \Omega).$$

Then $u \in C(E) \cap \mathcal{H}(E^\circ)$ by uniform convergence, but there does not exist v in $\mathcal{H}(E)$ such that $\left|v(X) - u(X)\right| < f\left(|X|\right)$ on E.

Details. To see this, we suppose that such a function v exists. Thus v is harmonic on some open set ω which contains E. Also, we observe from a closer examination of the final inequality in the proof of Lemma 5.1 that, if $a \geq n$, then we may take $C(n, a)$ to be 2^a in the statement of that result. Hence, if $|X| > 6$, we obtain

$$\left|v(X) - u(X)\right| \leq 2^a \left(3/|X|\right)^a f(3) \to 0 \qquad (a \to +\infty).$$

It follows that $v = u$ on $\omega \backslash (\cup_k B_k)$. This, in turn, implies that v fails to satisfy the spherical mean value equality on a neighbourhood of any B_k contained in ω. Since $B_k \subset \omega$ for all sufficiently large values of k, we obtain a contradiction.

Theorem 5.13. *The following are equivalent:*
(a) for each u in $C(E) \cap \mathcal{H}(E^\circ)$ and each positive number ϵ there exists v in $\mathcal{H}(\mathbf{R}^n)$ such that $|v - u| < \epsilon$ on E;
(b) for each u in $C(E) \cap \mathcal{H}(E^\circ)$ and each choice of positive numbers a and ϵ there exists v in $\mathcal{H}(\mathbf{R}^n)$ such that (5.17) holds;
(c) $\mathbf{R}^n \backslash \widehat{E}$ and $\mathbf{R}^n \backslash E^\circ$ are thin at the same points of E, and (Ω, E) satisfies the (K, L)-condition.

Proof. This follows immediately from Theorems 5.9 and 5.11.

Notes

Theorems 5.9 and 5.13 are new. All other results in this chapter (including Corollary 5.10) are due to Armitage and Goldstein [AG1]. The above theorems deal with pairs of the form $(\mathbf{R}^n, E)$, but the same methods apply to pairs of the form (Ω, E), where Ω is any unbounded open set in $\mathbf{R}^n$ (see [AG1]): in this case we obtain good approximation near one boundary point, namely ∞. It was also shown in [AG1] that, in Corollary 5.10, one can arrange for all partial derivatives of v, up to a predetermined order, to approximate the corresponding partial derivatives of u, with the same error as in (5.10).

strong extension pairs. A characterization of extension pairs will be given in §§6.3, 6.4.

Theorem 6.1. *The following are equivalent:*
(a) (Ω, E) is a strong extension pair for superharmonic functions;
(b) (Ω, E) is a strong extension pair for continuous superharmonic functions;
(c) $\Omega^*\backslash E$ is connected and locally connected.

The following lemma will be useful in the proof of Theorem 6.1 and later results.

Lemma 6.2. *Let ω be an open set and K be a compact subset of ω. Further, let $\omega_1, \omega_2, \ldots, \omega_m$ be the bounded components of $\mathbf{R}^n\backslash K$ which are not subsets of ω, and let $Y_k \in \omega_k$ $(k = 1, 2, \ldots, m)$. If $u \in \mathcal{S}(\omega)$ (resp. $u \in C(\omega) \cap \mathcal{S}(\omega)$), then there exists v in $\mathcal{S}(\mathbf{R}^n)$ (resp. $v \in C(\mathbf{R}^n) \cap \mathcal{S}(\mathbf{R}^n)$) and a positive constant c such that*

$$u(X) = v(X) - c\sum_{k=1}^{m} \phi_n(|X - Y_k|)$$

on some open set which contains K.

Proof of Lemma. Let $u \in \mathcal{S}(\omega)$ (resp. $u \in C(\omega) \cap \mathcal{S}(\omega)$). Without loss of generality we may assume that ω is bounded. Let ω_{m+1} denote the unbounded component of $\mathbf{R}^n\backslash K$. For each k in the set $\{1, 2, \ldots, m+1\}$ let U_k, V_k be open subsets of ω_k with Dirichlet regular Euclidean boundaries, such that U_k is connected and

$$\omega_k\backslash\omega \subset U_k, \quad \partial U_k \subset V_k \text{ and } \overline{V_k} \subset \omega \cap \omega_k.$$

If $k \in \{1, 2, \ldots, m\}$, then we further arrange that $Y_k \in U_k\backslash\overline{V_k}$. We define

$$U = \bigcup_{k=1}^{m+1} U_k, \qquad V = \bigcup_{k=1}^{m+1} V_k$$

and

$$w(X) = \begin{cases} H_u^V(X) & (X \in V) \\ u(X) & (X \in \omega\backslash V). \end{cases}$$

Thus $w \in \mathcal{S}(\omega)$ (resp. $w \in C(\omega) \cap \mathcal{S}(\omega)$) and $w \in \mathcal{H}(V)$. Next we define h on U as follows. On U_k $(k = 1, 2, \ldots, m)$ let h be the Green function for U_k with pole at Y_k. On U_{m+1} let h be the Green function for $U_{m+1} \cup \{\infty\}$ with

6 Superharmonic Extension and Approximation

6.1 Introduction

Many of the results in the preceding chapters have superharmonic analogues, some of which we will consider in this chapter. Thus, for example, we will examine which pairs (Ω, E) have the property that every u in $\mathcal{S}(E)$ can be uniformly approximated on E by functions in $\mathcal{S}(\Omega)$. However, in the case of superharmonic functions, it may be possible not only to approximate, but even to extend, functions in $\mathcal{S}(E)$.

For example, suppose that K is a compact subset of an open set Ω such that $\Omega^*\backslash K$ is connected. Then, as we saw in Theorem 1.7, for every u in $\mathcal{H}(K)$ and every positive number ϵ, there exists v in $\mathcal{H}(\Omega)$ such that $|v - u| < \epsilon$ on K. The corresponding fact for superharmonic functions (a special case of Theorem 6.1 below) is that, for every u in $\mathcal{S}(K)$ there exists v in $\mathcal{S}(\Omega)$ such that $v = u$ on K.

We begin the chapter with some such extension results. Later we will deal with Runge and Arakelyan approximation. Throughout this chapter Ω denotes an open set in $\mathbf{R}^n$ and E is a relatively closed subset of Ω. By a *continuous* superharmonic function we mean one which is both finite-valued and continuous.

6.2 Strong Extension

We call (Ω, E) an *extension pair for superharmonic functions* (resp. *for continuous superharmonic functions*) if, for each function (resp. each continuous function) u in $\mathcal{S}(E)$ there exists v in $\mathcal{S}(\Omega)$ (resp. in $C(\Omega) \cap \mathcal{S}(\Omega)$) such that $v = u$ on E. Further, (Ω, E) will be called a *strong extension pair for superharmonic functions* (resp. *for continuous superharmonic functions*) if it can be arranged that $v = u$ on an open set which contains E. In the latter case we preserve not only the values of u on E, but also the associated Riesz measure on an open set which contains E. In this section we characterize pole at ∞ if $n = 2$, or the solution to the Dirichlet problem on U_{m+1} with boundary data 0 on ∂U_{m+1} and 1 at ∞ if $n \geq 3$.

Now let

$$M > \sup\{w(X) : X \in \partial U\},$$

$$c > \sup\left(\left\{\frac{M - w(X)}{h(X)} : X \in U \cap \partial V\right\} \cup \{0\}\right)$$

and

$$s(X) = \begin{cases} w(X) & (X \in \omega \backslash U) \\ \min\{w(X), M - ch(X)\} & (X \in U \cap V) \\ M - ch(X) & (X \in U \backslash V). \end{cases}$$

It is straightforward to check that s is superharmonic (resp. continuous and superharmonic) on $\mathbf{R}^n \backslash \{Y_1, Y_2, \ldots, Y_m\}$, and clearly $s = u$ on an open set which contains K. Finally, the function v defined by

$$v(X) = s(X) + c\sum_{k=1}^{m} \phi_n(|X - Y_k|) \qquad (X \notin \{Y_1, Y_2, \ldots, Y_m\})$$

has a superharmonic (resp. continuous superharmonic) extension to all of $\mathbf{R}^n$. This completes the proof of the lemma.

Proof of Theorem 6.1. We begin by showing that (c) implies (a) (resp. (b)). A subset A of Ω will be called *Ω-solid* if $\Omega^* \backslash A$ is connected. Let u be a function (resp. continuous function) in $\mathcal{S}(E)$ and let $X_0 \in E$. Using Lemma 3.2 we can choose (L_m) to be an exhaustion of Ω such that $L_1 = \{X_0\}$ and such that, for each m, the sets L_m and $L_m \cup E$ are Ω-solid.

Next we inductively define a sequence (u_m) of functions (resp. continuous functions) such that, for each m,

(I) $u_m \in \mathcal{S}(L_m \cup E)$,

(II) $u_m = u$ on an open set U_m which contains E,

and such that $u_{m+1} = u_m$ on L_m. If we define $u_1 = u$, then (I) and (II) hold when $m = 1$. Given u_m, we construct u_{m+1} as follows. We know that u_m is superharmonic (resp. continuous and superharmonic) on an open subset ω of Ω such that $L_m \cup E \subset \omega$. Hence ω also contains the compact set K defined by $K = L_{m+2} \cap (L_m \cup E)$. Since, by the choice of (L_m), the sets L_{m+2} and $L_m \cup E$ are Ω-solid, it follows that K is Ω-solid. Thus, noting that $\operatorname{dist}(K, \mathbf{R}^n \backslash \Omega) > 0$, we see that $\mathbf{R}^n \backslash K$ has finitely many bounded components $\omega_1, \omega_2, \ldots, \omega_l$, and we can choose Y_k in $\omega_k \backslash \Omega$ for each k in $\{1, 2, \ldots, l\}$. Lemma 6.2 can now be applied (with u replaced by u_m) to obtain a superharmonic (resp. continuous superharmonic) function v_m on $\mathbf{R}^n \backslash \{Y_1, Y_2, \ldots, Y_l\}$, and hence on Ω, such that $v_m = u_m$ on an open set ω' which contains K. We define $V = (\omega \backslash L_{m+2}) \cup \omega'$ and

$$u_{m+1}(X) = \begin{cases} v_m(X) & (X \in L_{m+2}^\circ) \\ u_m(X) & (X \in V). \end{cases}$$

This function is well-defined, and hence superharmonic (resp. continuous and superharmonic) on the open set $L_{m+2}^\circ \cup V$, since the two parts of the definition agree on the region of overlap, namely $L_{m+2}^\circ \cap \omega'$. We know that

$$E \backslash L_{m+2} \subseteq \omega \backslash L_{m+2} \text{ and } E \cap L_{m+2} \subseteq K \subseteq \omega',$$

so $E \subseteq V$ and

$$L_{m+1} \cup E \subseteq L_{m+2}^\circ \cup V.$$

It follows that u_{m+1} is superharmonic (resp. continuous and superharmonic) on an open set containing $L_{m+1} \cup E$; that $u_{m+1} = u_m = u$ on the open set U_{m+1} defined by $U_{m+1} = U_m \cap V$, which contains E; and that $u_{m+1} = u_m$ on L_m, since $L_m \subseteq K \subset \omega' \subseteq V$.

The final step of this part of the proof is to define $v(X) = \lim_{m \to \infty} u_m(X)$ for each X in Ω. Given Y in Ω, there exists m' such that $Y \in L_{m'}^\circ$, and $u_m = u_{m'}$ on $L_{m'}$ whenever $m \geq m'$. It follows that v is superharmonic (resp. continuous and superharmonic) on a neighbourhood of Y, and thus, in view of the arbitrary nature of Y, on all of Ω. From property (II) above, and the fact that $v = u_m$ on L_m, it is clear that $v = u$ on the open set $\cup_m (U_m \cap L_m^\circ)$ which contains E. Hence (a) (resp. (b)) is established.

Conversely, suppose that (a) (resp. (b)) holds. If $\Omega^* \backslash E$ is not connected, then there is an Ω-bounded component V of $\Omega \backslash E$. We fix Y in V, define $u(X) = -\phi_n(|X - Y|)$ and conclude, by hypothesis, that there exists v in $\mathcal{S}(\Omega)$ such that $v = u$ on an open set ω which contains E. Now let W be an Ω-bounded connected open set such that $\overline{V} \subset W$ and $\overline{W} \subset \omega \cup V$. Since u is subharmonic on Ω, we know that $H_u^V \leq H_u^W$ on V. Since $v \in \mathcal{S}(\Omega)$, it is also true that $H_v^V \geq H_v^W$ on V. Observing that $v = u$ on ∂V and ∂W, it follows that $H_u^V = H_u^W$ on V. Hence $H_u^W - u$, which is a positive superharmonic function on W, takes the value 0 at every regular boundary point of V: a contradiction. Thus $\Omega^* \backslash E$ must be connected. It also follows by hypothesis and Theorem 3.14 that $\Omega^* \backslash \widehat{E}$ is locally connected. The connectedness of $\Omega^* \backslash E$, shown above, means that $\widehat{E} = E$, so $\Omega^* \backslash E$ is locally connected. This completes the proof of Theorem 6.1.

6.3 Extension from Compact Sets

The characterization of extension pairs involves a comparison of harmonic measures. Recall that, if V is a Greenian open set, then we use $\mu_{V,X}$ to denote harmonic measure for V and a point X in V. Also, we note that, if

K is a compact subset of Ω, then $\operatorname{dist}(\widehat{K}, \mathbf{R}^n\backslash\Omega) > 0$, and so $\mathbf{R}^n\backslash\widehat{K}$ has only finitely many components. (Recall that $\widehat{K}$ is defined with respect to Ω: it is the union of K with all Ω-bounded components of $\Omega\backslash K$.) The collection of all Borel subsets of $\mathbf{R}^n$ will be denoted by $\mathcal{B}$. A complete characterization of extension pairs will be given §6.4. In this section we present a special case of the solution which has a simpler formulation.

Theorem 6.3. *Let K be a compact subset of Ω such that $\widehat{K}$ is not thin at any of its points. Then (Ω, K) is an extension pair for superharmonic functions if and only if each Ω-bounded component V_0 of $\Omega\backslash K$ satisfies the following conditions:*
(i) V_0 is regular for the Dirichlet problem, and
(ii) given X_k in V_k $(k = 0, 1, \ldots, m)$, where $V_1, V_2, \ldots, V_m$ denote the components of $\mathbf{R}^n\backslash\widehat{K}$, there are positive constants $c_1, c_2, \ldots, c_m$ such that

$$\mu_{V_0,X_0}(A) \le \sum_{k=1}^{m} c_k \mu_{V_k,X_k}(A) \qquad (A \in \mathcal{B}). \tag{6.1}$$

It will be clear from the proof given below that the same characterization is valid for extension pairs for continuous superharmonic functions. We observe from Harnack's inequalities that, if there exist constants $c_1, c_2, \ldots, c_m$ such that (6.1) holds for a given choice of $X_0, X_1, \ldots, X_m$, then corresponding constants can be found for any other choice of $X_0, X_1, \ldots, X_m$. Also, conditions (i) and (ii) above together imply that $\partial V_0 \subseteq \partial\widehat{K}$. To see this, we note that $\partial V_0\backslash\partial\widehat{K}$ is a relatively open subset of ∂V_0 which has zero harmonic measure for V_0, by (ii). Thus every point of $\partial V_0\backslash\partial\widehat{K}$ is irregular, and (i) now shows that $\partial V_0 \subseteq \partial\widehat{K}$.

Before proving Theorem 6.3, we will illustrate it by means of several pairs of the form $(\mathbf{R}^2, K)$, where K is the union of finitely many line segments. Our assertions are based on the observation that, if

$$S_\alpha = \{re^{i\theta} : 0 < \theta < \alpha \text{ and } 0 < r < 2\} \text{ and } z_\alpha = e^{i\alpha/2} \quad (0 < \alpha \le 2\pi),$$

then (identifying $\mathbf{R}^2$ with $\mathbf{C}$ in the usual manner) the restriction of μ_{S_α,z_α} to the interval (0,2) is absolutely continuous with respect to one-dimensional Lebesgue measure λ', and there are positive constants $C_1(\alpha), C_2(\alpha)$ such that

$$C_1(\alpha)t^{\pi/\alpha-1} \le (d\mu_{S_\alpha,z_\alpha}/d\lambda')(t) \le C_2(\alpha)t^{\pi/\alpha-1} \qquad (0 < t \le 1).$$

Example 6.4.
(a) Let P denote an open polygon in $\mathbf{R}^2$. Then $(\mathbf{R}^2, \partial P)$ is an extension pair for superharmonic functions if and only if P is convex.

(b) Let

$$A_1 = [0,2]^2 \cup ([2,4] \times \{2\}), \quad A_2 = [0,2]^2 \cup ([2,4] \times \{1\}),$$
$$A_3 = ([0,1] \cup [2,3])^2 \cup [1,2]^2, \quad A_4 = ([0,1) \cup (1,2]) \times [0,1],$$
$$A_5 = [0,2]^2 \backslash \{(1,1)\}$$

(see Figure 6.1). Then $(\mathbf{R}^2, \partial A_1)$ and $(\mathbf{R}^2, \partial A_3)$ are extension pairs for superharmonic functions. However $(\mathbf{R}^2, \partial A_2)$ and $(\mathbf{R}^2, \partial A_4)$ violate condition (ii) of Theorem 6.3, and $(\mathbf{R}^2, \partial A_5)$ violates condition (i), so these are not extension pairs.

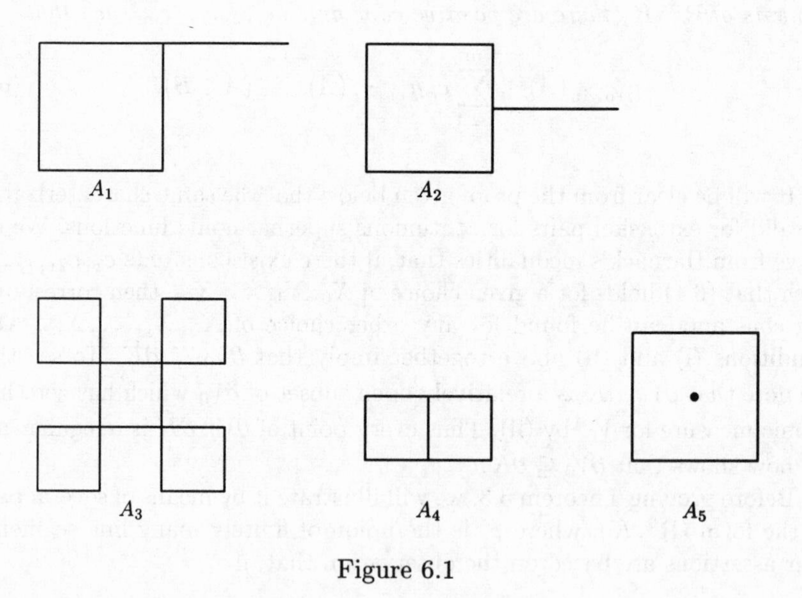

Figure 6.1

Proof of Theorem 6.3. Let Ω and K be as in the first sentence of Theorem 6.3, suppose that each Ω-bounded component V_0 of $\Omega \backslash K$ satisfies conditions (i) and (ii) of the theorem, and let u be a superharmonic function on some open set ω (where $\omega \subseteq \Omega$) which contains K. Further, let $U_1, U_2, \ldots, U_l$ be the bounded components of $\mathbf{R}^n \backslash K$ which are not subsets of Ω, and let $W_1, W_2, \ldots, W_p$ be the remaining bounded components of $\mathbf{R}^n \backslash K$ which are not subsets of ω. We choose Y_k in $U_k \backslash \Omega$ for each k in $\{1, 2, \ldots, l\}$, and Z_k

in W_k for each k in $\{1,2,\ldots,p\}$. It follows from Lemma 6.2 that there is a positive constant c and a superharmonic function v' on $\mathbf{R}^n$ such that

$$u(X) = v'(X) - c\left\{\sum_{k=1}^{l}\phi_n(|X-Y_k|) + \sum_{k=1}^{p}\phi_n(|X-Z_k|)\right\} \quad (X \in E).$$

In particular, there is a superharmonic function v_1 on Ω such that

$$u(X) = v_1(X) - c\sum_{k=1}^{p}\phi_n(|X-Z_k|) \qquad (X \in E). \tag{6.2}$$

Now let $V_0 = W_1$ and $X_0 = Z_1$. Further, let $V_1, V_2, \ldots, V_m$ denote the components of $\mathbf{R}^n \backslash \widehat{K}$ and let $X_k \in V_k$ $(k = 1,2,\ldots,m)$. By hypothesis (ii) there are positive constants $c_1, c_2, \ldots, c_m$ such that (6.1) holds. We now define

$$\begin{aligned} s_1(X) &= \phi_n(|X-X_0|) - \int \phi_n(|X-Y|)\, d\mu_{V_0,X_0}(Y) \\ &\quad + \sum_{k=1}^{m} c_k\left\{\int \phi_n(|X-Y|)\, d\mu_{V_k,X_k}(Y) - \phi_n(|X-X_k|)\right\}. \end{aligned} \tag{6.3}$$

Inequality (6.1) shows that the function $s_1(X) - \phi_n(|X-Z_1|)$ is superharmonic on $\mathbf{R}^n\backslash\{X_1, X_2, \ldots, X_m\}$, in view of the fact that $X_0 = Z_1$. Further, the regularity of V_0 (hypothesis (i)) and the fact that $\widehat{K}$ is not thin at any of its points (first sentence of Theorem 6.3) ensure that $s_1 = 0$ on K. To see this when $n \geq 3$ we note that, if $X \in K$, then $\mathbf{R}^n \backslash V_k$ is not thin at X, and so Theorems 0.J and 0.I yield

$$\begin{aligned} \phi_n(|X-X_k|) &= R^{\mathbf{R}^n\backslash V_k}_{\phi_n(|X-\cdot|)}(X_k) \\ &= \int \phi_n(|X-Y|)\, d\mu_{V_k,X_k}(Y) \quad (k = 0,1,\ldots,m), \end{aligned}$$

where reductions are with respect to superharmonic functions on $\mathbf{R}^n$. A modified form of this argument applies also when $n = 2$. Hence, if we define

$$v_2(X) = v_1(X) + c\left\{s_1(X) - \phi_n(|X-Z_1|)\right\},$$

we obtain a superharmonic function v_2 on $\Omega\backslash\{X_1, X_2, \ldots, X_m\}$ such that

$$u(X) = v_2(X) - c\sum_{k=2}^{p}\phi_n(|X-Z_k|) \qquad (X \in K)$$

(see (6.2)).

If we repeat the argument of the preceding paragraph with the assignments $V_0 = W_k$ and $X_0 = Z_k$ $(k = 2, 3, \ldots, p)$, we obtain a superharmonic function v_{p+1} on $\Omega\backslash\{X_1, X_2, \ldots, X_m\}$ such that $u = v_{p+1}$ on K. Since $v_{p+1} \in \mathcal{S}(\widehat{K})$ and the set $\Omega^*\backslash\widehat{K}$ is connected, we can apply Theorem 6.1 to the pair $(\Omega, \widehat{K})$ to conclude that there is a superharmonic function v on Ω such that $v = v_{p+1}$ on $\widehat{K}$, and hence $v = u$ on K. It follows that (Ω, K) is an extension pair for superharmonic functions.

Conversely, suppose that (Ω, K) is an extension pair for superharmonic functions, let V_0 be an Ω-bounded component of $\Omega\backslash E$, let $X_0 \in V_0$, and define $u(X) = -\phi_n(|X - X_0|)$. By hypothesis there is a superharmonic function v on Ω such that $v = u$ on K. Thus the function defined by $w = v - u$ is superharmonic on Ω and vanishes on ∂V_0. In particular, w is a positive superharmonic function on V_0 which vanishes on ∂V_0, so V_0 must be regular for the Dirichlet problem and (i) is proved.

Now let $V_1, V_2, \ldots, V_m$ be the components of $\mathbf{R}^n\backslash\widehat{K}$, and let $G_k(\,\cdot\,,\,\cdot\,)$ be the Green function for V_k for each k in $\{0, 1, \ldots, m\}$. Further, let W be an Ω-bounded open set which contains $\widehat{K}$ and let $X_k \in V_k\backslash\overline{W}$ for each k in $\{1, 2, \ldots, m\}$. For each k in the latter set we can find a positive constant c_k such that

$$-c_k G_k(X_k, X) < w(X) \qquad (X \in \partial W \cap V_k), \tag{6.4}$$

where w is as in the previous paragraph. It follows from the minimum principle that the inequality in (6.4) remains true for all X in $W \cap V_k$. It is also clear that $w \geq G_0(X_0, \,\cdot\,)$ on V_0 and that $w \geq 0$ on $\widehat{K}$. Hence the function s defined on $\mathbf{R}^n$ by

$$s(X) = \begin{cases} G_0(X_0, X) & (X \in V_0) \\ -c_k G_k(X_k, X) & (X \in V_k; k \in \{1, 2, \ldots, m\}) \\ 0 & (X \in \widehat{K}\backslash V_0) \end{cases}$$

is superharmonic on $\mathbf{R}^n\backslash\{X_1, X_2, \ldots, X_m\}$. The function s can also be written as in (6.3). Since $\Delta s \leq 0$ on $\mathbf{R}^n\backslash\{X_1, X_2, \ldots, X_m\}$ in the sense of distributions, we conclude that (6.1) holds. Thus Theorem 6.3 is established.

6.4 Extension from Relatively Closed Sets

We come now to the question of characterizing extension pairs (Ω, E) in the absence of any special conditions on E. If W is an open set which satisfies $\widehat{E} \subseteq W \subseteq \Omega$, then we define a class of superharmonic functions on W by

$$\mathcal{S}_W = \{w \in \mathcal{S}^+(W) : w = 1 \text{ on } \widehat{E}\}.$$

Also, the Riesz measure associated with a superharmonic function w is denoted by ν_w. As usual, E denotes a relatively closed subset of Ω. By a countable set we mean one which is either finite or countably infinite.

Theorem 6.5. *The pair (Ω, E) is an extension pair for superharmonic functions if and only if:*
(i) each Ω-bounded component of $\Omega\backslash E$ is regular for the Dirichlet problem,
(ii) $\Omega^*\backslash\widehat{E}$ is locally connected, and
(iii) for each countable collection $\{(X_k, c_k) : k \in I\}$ of pairs from $(\widehat{E}\backslash E) \times (0, +\infty)$ such that the points X_k are distinct and have no limit point in Ω, there exists an open set W satisfying $\widehat{E} \subseteq W \subseteq \Omega$ and a function w in $\mathcal{S}_W$ such that

$$\sum_{k\in I} c_k \mu_{(\widehat{E}\backslash E), X_k}(A) \leq \nu_w(A) \qquad (A \in \mathcal{B}). \tag{6.5}$$

As in §6.3 we observe that conditions (i) and (iii) above together imply that $\partial V \subseteq \partial\widehat{E}$ for each Ω-bounded component V of $\Omega\backslash E$. Condition (iii) is similar in nature to condition (ii) of Theorem 6.3 but, together with (ii), it also implies that (Ω, E) satisfies the (K, L)-condition: see Lemma 6.7 below. First we give an example.

Example 6.6. Let (P_k) be a sequence of open polygons in $\mathbf{R}^2$ such that the closures $\overline{P_k}$ are pairwise disjoint, and such that only a finite number of the polygons intersect any given compact set. Then $(\mathbf{R}^2, \cup_k \partial P_k)$ is an extension pair for superharmonic functions if and only if each of the polygons P_k is convex. The "if" part of this assertion can be checked using the observations preceding Example 6.4: we choose W to be $\cup_k Q_k$ in condition (iii) of Theorem 6.5, where (Q_k) is a suitable sequence of pairwise disjoint open sets such that $\overline{P_k} \subset Q_k$ for each k. The "only if" part follows from Example 6.4(a).

Before proving Theorem 6.5 we establish the following lemma.

Lemma 6.7. *Suppose that (Ω, E) satisfies condition (iii) of Theorem 6.5. Then, for each compact subset K of Ω, there is a compact subset L of Ω which contains every Ω-bounded component of $\Omega\backslash E$ that intersects K.*

Proof of Lemma. Suppose that condition (iii) of Theorem 6.5 holds, but that the conclusion of the lemma fails. Then there exist a compact subset K of Ω, a sequence (V_k) of distinct Ω-bounded components of $\Omega\backslash E$, and sequences $(X_k), (Y_k)$ of points, such that $X_k, Y_k \in V_k$ for each k, and such that $X_k \to \mathcal{A}$ and (Y_k) converges to some point Y_0 in K. Now let U be an Ω-bounded open set which contains K and let U_0 be the component of U which contains Y_0, define

$$a_k = \mu_{V_k, X_k}(U_0 \cap \partial V_k) \qquad (k \in \mathbf{N}),$$

and let $c_k = a_k^{-1}$. (We know that $a_k > 0$: see the first paragraph of the proof of Theorem 3.14.) Inequality (6.5) now implies that $\nu_w(\overline{U} \cap E) = +\infty$. This is impossible, since $\overline{U} \cap E$ is a compact subset of W. Hence the lemma is proved.

Proof of Theorem 6.5. Suppose that conditions (i)–(iii) of the theorem hold and let u be a superharmonic function on some open set ω (where $\omega \subseteq \Omega$) which contains E. We denote by $\{V_k : k \in I\}$ the collection of Ω-bounded components of $\Omega \backslash E$ which are not subsets of ω, choose X_k in V_k for each k in I, and let

$$\omega_1 = \omega \cup \left(\bigcup_{k \in I} (V_k \backslash \{X_k\}) \right).$$

Let K be a compact subset of Ω, and define

$$S = \bigcup_{k \in J} V_k, \text{ where } J = \{k \in I : \overline{V_k} \cap K \neq \emptyset\}.$$

It follows from (iii) and Lemma 6.7 that S is Ω-bounded. Since

$$\mathrm{dist}(\overline{S} \cap E, \mathbf{R}^n \backslash \omega) > 0$$

it is clear that J is a finite set. Thus $\{X_k : k \in I\}$ has no limit point in Ω. Next, for each k in I, we apply Lemma 6.2 with ∂V_k in place of E. This allows us to construct a superharmonic function s on ω_1 such that $s = u$ on an open set which contains E, and such that the function

$$s(X) + c_k \phi_n(|X - X_k|)$$

has a superharmonic extension to $\omega_1 \cup \{X_k\}$ for a suitable choice of positive constant c_k.

By condition (iii) there exist an open set W satisfying $\widehat{E} \subseteq W \subseteq \Omega$ and a function w in $\mathcal{S}_W$ such that (6.5) holds. Let

$$w'(X) = \sum_{k \in I} c_k \left\{ \phi_n(|X - X_k|) - \int \phi_n(|X - Y|)\, d\mu_{V_k, X_k}(Y) \right\}$$
$$+ s(X) + w(X) - 1 \qquad (X \in W \cap \omega_1).$$

Inequality (6.5) ensures that w is superharmonic on $W \cap \omega_1$, and we have arranged s in such a way that w' has a superharmonic extension to the set $W \cap (\omega \cup (\cup_k V_k))$, which contains $\widehat{E}$. Since $\Omega^* \backslash \widehat{E}$ is connected (by the definition of $\widehat{E}$) and locally connected (by (ii)), we can apply Theorem 6.1 to the pair $(\Omega, \widehat{E})$ to obtain v in $\mathcal{S}(\Omega)$ such that $v = w'$ on $\widehat{E}$. Also, $w' = s = u$ on E by condition (i) (see the argument given in the second paragraph of the proof of Theorem 6.3) and the definition of $\mathcal{S}_W$. Hence $v = u$ on E. It follows that (Ω, E) is an extension pair for superharmonic functions.

Conversely, suppose that (Ω, E) is an extension pair for superharmonic functions. It follows as in the proof of Theorem 6.3 that (i) holds. Also, Theorem 3.14 shows that (ii) holds. Thus it remains to establish (iii). Let $\{(X_k, c_k) : k \in I\}$ be a countable collection of pairs from $(\widehat{E} \backslash E) \times (0, +\infty)$ such that the points X_k are distinct and $\{X_k : k \in I\}$ has no limit point in Ω, and let $\Omega_1 = \Omega \backslash \{X_k : k \in I\}$. Using Corollary 3.9 we can choose u to be a harmonic function on Ω_1 such that $u(X) + c_k \phi_n(|X - X_k|)$ has a harmonic extension to $\Omega_1 \cup \{X_k\}$, for each k in I. By hypothesis there is a superharmonic function v on Ω such that $v = u$ on E. Since $v - u \in \mathcal{S}(\Omega)$, it follows from the minimum principle that $v - u \geq 0$ on $\widehat{E}$. For each k in I, let V_k be the component of $\widehat{E} \backslash E$ to which X_k belongs. We know from Lemma 6.7 that any given compact subset of Ω intersects only finitely many of the sets V_k. Also, let $G_k(\cdot, \cdot)$ be the Green function for V_k, and define $G_k(\cdot, \cdot) = 0$ outside $V_k \times V_k$. Clearly

$$v(X) - u(X) \geq \sum_{k \in I} c_k G_k(X_k, X) \qquad (X \in \cup_k V_k). \tag{6.6}$$

Let

$$W = \{X \in \Omega : v(X) - u(X) > -1\}$$

and

$$w(X) = 1 + \min\{v(X) - u(X), 0\} \qquad (X \in W).$$

Clearly W is an open set satisfying $\widehat{E} \subseteq W \subseteq \Omega$, and also $w \in \mathcal{S}_W$. Further, the function s defined by

$$s(X) = w(X) - 1 + \sum_{k \in I} c_k G_k(X_k, X) \qquad (X \in W)$$

is also superharmonic on W, in view of (6.6). We can rewrite s as

$$s(X) = w(X) - 1 + \sum_{k \in I} c_k \left\{ \phi_n(|X - X_k|) - \int \phi_n(|X - Y|)\, d\mu_{V_k, X_k}(Y) \right\}$$

on W, and so (6.3) must hold. This completes the proof of Theorem 6.5.

6.5 Runge Approximation

We now turn from extension results to approximation results, and give the following analogue of Theorem 3.15.

Theorem 6.8. *The following are equivalent:*
(a) for each u in $\mathcal{S}(E)$ and each positive number ϵ there exists v in $\mathcal{S}(\Omega)$ such that $u-\epsilon \le v \le u+\epsilon$ on E;
(b) for each u in $\mathcal{S}(E)$ and each s in $\mathcal{S}^+(E)$ there exists v in $\mathcal{S}(\Omega)$ such that $u \le v \le u+s$ on E;
(c) $\Omega\backslash\widehat{E}$ and $\Omega\backslash E$ are thin at the same points of E, and (Ω, E) satisfies the (K,L)-condition.

Again we remark that Theorem 6.8 remains true (with the same proof as given below) if the functions u and v are required to be continuous in (a) and (b) above.

Proof. Suppose that (c) holds, let $u \in \mathcal{S}(E)$ and $s \in \mathcal{S}^+(E)$. In view of Lemma 3.16 we may assume that $s \in \mathcal{S}^+(\widehat{E})$. There is an open subset ω of Ω such that $u \in \mathcal{S}(\omega)$ and $E \subset \omega$. We denote by $\{V_k : k \in I\}$ the collection of Ω-bounded components of $\Omega\backslash E$ which are not subsets of ω, choose Y_k in V_k for each k in I, and define

$$\omega_1 = \omega \cup \left(\bigcup_{k\in I} (V_k\backslash\{Y_k\}) \right).$$

For each k in I we apply Lemma 6.2 with ∂V_k in place of E. This allows us to construct a superharmonic function u' on ω_1 such that $u' = u$ on an open set which contains E, and such that the function $u'(X) + b_k\phi_n(|X - Y_k|)$ has a superharmonic extension to $\omega_1 \cup \{Y_k\}$ for a suitable choice of positive constant b_k, for each k. It was shown in the proof of Theorem 3.15 that there is an open subset W of Ω which contains $\widehat{E}$ and which has the following property: the equation

$$v_1(X) = \sum_{k\in I} b_k G_W(Y_k, X) \qquad (X \in W)$$

defines a potential on W and $v_1 \le s/2$ on E. (Here, as usual, G_W denotes the Green function for W.) Thus, if we define $v_2 = u' + v_1$, we see that v_2 has a superharmonic extension to the set $W \cap (\omega \cup (\cup_k V_k))$, which contains $\widehat{E}$. Since $\Omega^*\backslash\widehat{E}$ is connected and (by the (K,L)-condition) locally connected, we can apply Theorem 6.1 to obtain v in $\mathcal{S}(\Omega)$ such that $v = v_2$ on $\widehat{E}$. Thus $u \le v \le u+s$ on E, and we have established (b).

Clearly (b) implies (a).

Now suppose that (a) holds. It follows from Theorem 3.14 that (Ω, E) satisfies the (K, L)-condition. Let $u \in \mathcal{H}(E)$ and $\epsilon > 0$. Then there exist v_1, v_2 in $\mathcal{S}(\Omega)$ such that

$$|v_1 - u| < \epsilon/2 \text{ and } |v_2 - (-u)| < \epsilon/2 \text{ on } E.$$

Thus the open set

$$U = \{X \in \Omega : v_1(X) + v_2(X) > -\epsilon\}$$

must contain not only E, but also $\widehat{E}$ in view of the minimum principle. Since $-v_2 - \epsilon/2$ is a subharmonic minorant of the superharmonic function $v_1 + \epsilon/2$ on U, the greatest harmonic minorant h must satisfy

$$-v_2 - \epsilon/2 \le h \le v_1 + \epsilon/2 \text{ on } W,$$

whence $|h - u| < \epsilon$ on E. Since $h \in \mathcal{H}(\widehat{E})$, we can now apply Theorem 1.14 to deduce that $\Omega\backslash\widehat{E}$ and $\Omega\backslash E$ are thin at the same points of E.

6.6 Approximation on Compact Sets

Let K be a compact subset of Ω. In this section we deal with uniform approximation of functions in $C(K) \cap \mathcal{S}(K^\circ)$, firstly by continuous functions in $\mathcal{S}(K)$, and later by functions in $C(\Omega) \cap \mathcal{S}(\Omega)$. The following is a superharmonic analogue of Theorem 1.3.

Theorem 6.9. *Let K be a compact subset of $\mathbf{R}^n$. The following are equivalent:*
(a) for each u in $C(K) \cap \mathcal{S}(K^\circ)$ and each positive number ϵ there exists a continuous function v in $\mathcal{S}(K)$ such that $|v - u| < \epsilon$ on K;
(b) $\mathbf{R}^n\backslash K$ and $\mathbf{R}^n\backslash K^\circ$ are thin at the same points of K.

Proof. Let K be a compact set such that (b) holds, let $(W(m))$ be a decreasing sequence of regular open sets which satisfy

$$K \subset W(m) \subset \{X : \text{dist}(X, K) < 1/m\} \qquad (m \in \mathbf{N}),$$

and let (L_k) be an exhaustion of K° such that L_k is not thin at any of its points. Also let $\epsilon > 0$ and $u \in C(K) \cap \mathcal{S}(K^\circ)$. By Tietze's extension theorem we can obtain $\tilde{u}$ in $C(\mathbf{R}^n)$ such that $\tilde{u} = u$ on K. We choose v_1, k' and m_1 as in the second paragraph of §1.4.

It follows from Lemma 1.5 (case (i)) that

$$H_{\tilde{u}}^{W(m)}(X) \to H_u^{K^\circ}(X) \qquad (m \to \infty; X \in K^\circ), \tag{6.7}$$

and that

$$\begin{aligned} G_{W(m)}(X,Y) &= \phi_n(|X-Y|) - H^{W(m)}_{\phi_n(|X-\cdot|)}(Y) \\ &\downarrow \phi_n(|X-Y|) - H^{K^\circ}_{\phi_n(|X-\cdot|)}(Y) \\ &= G_{K^\circ}(X,Y) \qquad (X,Y \in K^\circ), \end{aligned} \tag{6.8}$$

where $G_{W(m)}$ and G_{K° denote the Green functions for $W(m)$ and K° respectively. (The function $\phi_n(|X-\cdot|)$ can be suitably truncated near X to make it finite and continuous on $\mathbf{R}^n$ without changing its values on $\mathbf{R}^n \backslash K^\circ$.) Let ν_u be the Riesz measure associated with u on K°. Thus $u = u' + u''$ on K°, where $u'(X) = H_u^{K^\circ}$ and

$$u''(X) = \int_{K^\circ} G_{K^\circ}(Y,X)\, d\nu_u(Y) \qquad (X \in K^\circ).$$

The convergence in (6.7) is locally uniform on K°, so we can choose m_2 such that $m_2 \geq m_1$ and

$$\left|H_{\tilde{u}}^{W(m)}(X) - u'(X)\right| < \epsilon/6 \qquad (X \in L_{k'}; m \geq m_2). \tag{6.9}$$

By Dini's Theorem we can choose l such that $l > k'$ and

$$\int_{L_l} G_{K^\circ}(Y,X)\, d\nu_u(Y) > u''(X) - \epsilon/6 \qquad (X \in L_{k'}).$$

Also, in view of (6.8), we can choose m_3 such that $m_3 \geq m_2$ and

$$w(X) < \int_{L_l} G_{K^\circ}(Y,X)\, d\nu_u(Y) + \epsilon/6 \qquad (X \in L_{k'}),$$

where

$$w(X) = \int_{L_l} G_{W(m_3)}(Y,X)\, d\nu_u(Y) \qquad (X \in W(m_3))$$

and $w = 0$ on $\partial W(m_3)$. Thus

$$\left|w(X) - u''(X)\right| < \epsilon/6 \qquad (X \in L_{k'}). \tag{6.10}$$

We now define the continuous superharmonic function v on $W(m_3)$ by

$$v(X) = \begin{cases} h(X) + H_w^{W(m_3)\backslash L_{k'}}(X) & (X \in W(m_3)\backslash L_{k'}) \\ h(X) + w(X) & (X \in L_{k'} \cup \partial W(m_3)), \end{cases}$$

where

$$h(X) = \begin{cases} H_{\tilde{u}}^{W(m_3)}(X) & (X \in W(m_3)) \\ \tilde{u}(X) & (X \in \partial W(m_3)). \end{cases}$$

It follows from (6.9) and (6.10) that

$$|v - u| \le |h - u'| + |w - u''| < \epsilon/3 \text{ on } L_{k'}. \tag{6.11}$$

Further,

$$|v - \tilde{u}| < \epsilon \text{ on } W(m_3)\backslash L_{k'} \tag{6.12}$$

by exactly the same reasoning as was used to obtain (1.13). Combining (6.11) and (6.12), we obtain $|v - u| < \epsilon$ on K, which establishes (a).

Conversely, suppose that (a) holds, let $u \in C(K) \cap \mathcal{H}(K^\circ)$ and $\epsilon > 0$. Applying (a) to the functions u and $-u$ we obtain continuous functions v_1, v_2 in $\mathcal{S}(K)$ such that

$$u < v_1 < u + \epsilon \text{ and } -u < v_2 < -u + \epsilon \text{ on } K.$$

Thus $-v_2 < v_1$ on some open set W which contains K. Since $-v_2$ is a subharmonic minorant of the superharmonic function v_1 on W, there exists a greatest harmonic minorant v of v_1 on W which satisfies $-v_2 \le v \le v_1$ on W and hence $|v - u| < \epsilon$ on K. It now follows from Theorem 1.6 that (b) holds. This completes the proof of Theorem 6.9.

Theorem 6.10. *Let K be a compact subset of an open set Ω. The following are equivalent:*
(a) for each u in $C(K) \cap \mathcal{S}(K^\circ)$ and each positive number ϵ there exists v in $C(\Omega) \cap \mathcal{S}(\Omega)$ such that $|v - u| < \epsilon$ on K;
(b) $\mathbf{R}^n \backslash \widehat{K}$ and $\mathbf{R}^n \backslash K^\circ$ are thin at the same points of K.

Proof. This can be proved by combining Theorems 6.8 and 6.9. Alternatively, it may be deduced from Theorem 6.9 by imitating the more direct argument presented for Theorem 1.15. In the latter case we use Theorem 6.1 in place of Theorem 1.7.

6.7 Approximation of Functions in $C(E) \cap \mathcal{S}(E^\circ)$

We will now obtain superharmonic analogues of Theorems 3.17 and 3.19.

Theorem 6.11. *The following are equivalent:*
(a) for each u in $C(E) \cap \mathcal{S}(E^\circ)$ and each positive number ϵ there exists a

continuous function v in $\mathcal{S}(E)$ such that $|v-u|<\epsilon$ on E;
(b) for each u and s in $C(E)\cap\mathcal{S}(E^\circ)$, where $s>0$, there exists a continuous function v in $\mathcal{S}(E)$ such that $0<v-u<s$ on E;
(c) $\Omega\backslash E$ and $\Omega\backslash E^\circ$ are thin at the same points of E.

Proof. Clearly (b) implies (a). If (a) holds, then (c) follows by the argument presented in the last paragraph of the proof of Theorem 6.9. It remains to prove that (c) implies (b). We will do this by first showing that (c) implies the following condition:

(b′) for each u in $C(E)\cap\mathcal{S}(E^\circ)$ and each s in $\mathcal{S}^+(E)$ there exists a continuous function v in $\mathcal{S}(E)$ such that $0<v-u<s$ on E.

Suppose that (c) holds, let $u\in C(E)\cap\mathcal{S}(E^\circ)$, and let s be a positive superharmonic function on an open set W which contains E. We may assume that W is Greenian (unless $E=\mathbf{R}^2$, in which case (b) trivially holds). Let $\{B_k : k\in\mathbf{N}\}$ be a covering of W by a locally finite collection of open balls such that $\overline{B_k}\subset W$ for each k. For each k let s_k be a potential on W which belongs to $C^\infty(W)$, and satisfies $\Delta s_k<0$ on B_k, $\Delta s_k=0$ on $W\backslash B_k$ and $s_k\le s$ on B_k. We define $s'=\sum_k 2^{-k}s_k$ on W. This new function is smooth and satisfies $\Delta s'<0$ and $0<s'\le s$ on W.

Now let (K_m) be an exhaustion of W such that $K_4\cap E=\emptyset$ and such that neither $W\backslash K_m$ nor K_m° is thin at any point of ∂K_m. We define $L_m=K_{m+2}\backslash K_m^\circ$ for each m in $\mathbf{N}$, and note from condition (c) that $W\backslash(L_m\cap E)$ and $W\backslash(L_m\cap E)^\circ$ must then be thin at the same points. For each m we choose ψ_m in $C^\infty(\mathbf{R}^n)$ such that $0\le\psi_m\le 1$ on $\mathbf{R}^n$, $\psi_m=0$ on an open set containing ∂K_{m+1}, and $\psi_m=1$ on $\mathbf{R}^n\backslash L_m^\circ$. Next, let (λ_m) be a decreasing sequence of positive numbers such that

$$\lambda_m<3^{-1}\inf\{s'(X): X\in K_{m+2}\} \tag{6.13}$$

and such that $(s'/3)+\lambda_m\psi_m$ is superharmonic on some W-bounded open set V_m which contains L_m. For each m we can apply Theorem 6.9 to obtain a continuous superharmonic function v_m on an open set U_m such that

$$L_m\cap E\subset U_m\subset\overline{U_m}\subset V_m$$

and

$$0<v_m(X)-u(X)<\lambda_{m+2}\qquad (X\in L_m\cap E). \tag{6.14}$$

We define the superharmonic function

$$f_m(X)=v_m(X)+3^{-1}s'(X)+\lambda_m\psi_m(X)\qquad (X\in U_m)$$

and the function

$$v(X) = \min\{f_{m-1}(X), f_m(X)\} \quad (X \in (K_{m+1} \backslash K_m) \cap U_{m-1} \cap U_m; m \geq 4).$$

If $X \in E \cap (K_{m+1} \backslash K_m)$, then

$$\begin{aligned} 0 < v(X) - u(X) &\leq \min\{v_{m-1}(X) - u(X), v_m(X) - u(X)\} \\ &\quad + 3^{-1}s'(X) + \max\{\lambda_{m-1}, \lambda_m\} \\ &< \lambda_{m+2} + 3^{-1}s'(X) + \lambda_{m-1} < s'(X), \end{aligned}$$

by (6.13) and (6.14). Hence $0 < v - u < s$ on E.

It remains to check that v is continuous and superharmonic on a neighbourhood of any point X_0 in E. This is clear from the definition if $X_0 \notin \cup_m \partial K_m$. If $X_0 \in \partial K_{m+1} \cap E$ and $k \in \{m-1, m+1\}$, then

$$f_m(X_0) - f_k(X_0) = v_m(X_0) - v_k(X_0) - \lambda_k.$$

Since

$$\left|v_m(X_0) - v_k(X_0)\right| < \max\{\lambda_{m+2}, \lambda_{k+2}\} \leq \lambda_{m+1} \leq \lambda_k,$$

by (6.14), it follows that $f_m(X_0) < f_k(X_0)$, and hence $f_m < f_k$ on a neighbourhood of X_0. Thus $v = f_m$ on a neighbourhood of X_0, which implies that v is continuous and superharmonic there. We have now shown that (c) implies (b′).

In order to complete the proof that (c) implies (b) we observe the following.

Lemma 6.12. *If condition (c) of Theorem 6.11 holds and $s \in C(E) \cap \mathcal{S}(E^{\circ})$, where $s > 0$, then there exists w in $\mathcal{S}(E)$ such that $0 < w < s$ on E.*

Proof of Lemma. This is directly analogous to the argument given for Lemma 3.18, and uses the fact, established above, that (c) implies (b′).

Thus, if (c) holds, we use Lemma 6.12 and the fact that (c) implies (b′) to conclude that (b) must hold. Theorem 6.11 is now established.

Theorem 6.13. *The following are equivalent:*
(a) for each u in $C(E) \cap \mathcal{S}(E^{\circ})$ and each positive number ϵ there exists v in $C(\Omega) \cap \mathcal{S}(\Omega)$ such that $|v - u| < \epsilon$ on E;
(b) for each u and s in $C(E) \cap \mathcal{S}(E^{\circ})$, where $s > 0$, there exists v in $C(\Omega) \cap \mathcal{S}(\Omega)$ such that $0 < v - u < s$ on E;
(c) $\Omega \backslash \widehat{E}$ and $\Omega \backslash E^{\circ}$ are thin at the same points of E, and (Ω, E) satisfies the (K, L)-condition.

Proof. Suppose that (c) holds, and let $u, s \in C(E) \cap \mathcal{S}(E^{\circ})$, where $s > 0$. Then condition (c) of Theorem 6.11 holds, so there exists a continuous function u' in $\mathcal{S}(E)$ such that $0 < u' - u < s/2$ on E. Also, by Lemma 6.12, there exists w in $\mathcal{S}^+(E)$ such that $0 < w < s$ on E. Since condition (c) of Theorem 6.8 also holds, we obtain v in $C(\Omega) \cap \mathcal{S}(\Omega)$ (see the remark following that result) such that $0 < v - u' < w/2$ on E. Hence $0 < v - u < s$ on E, proving (b).

Clearly (b) implies (a).

Finally, if (a) holds, then it follows from Theorems 6.8 and 6.11 that (c) must hold.

Notes

Theorem 6.1 was obtained independently by Gauthier [Gau3] and Gardiner [Gar3], though the formulation in [Gau3] is somewhat different from that presented here. Earlier extension results can be found in [Arm1]. Theorems 6.3 and 6.5 come from [Gar3], as does the equivalence of (a) and (c) in Theorem 6.8. Theorem 6.9 is contained in work of Bliedtner and Hansen [BH] in an abstract setting (see also Hansen [Han]). It also appears in [Sir], although the proof given there is incomplete. The particular case of Theorem 6.8 where E is assumed to be compact, is contained in [BH] (see also [GGG1]). Theorem 6.11 is a refinement of work of Gauthier [Gau2] (see also [BeG]). Superharmonic analogues of the results of Chapter 4 are now easy to obtain. A superharmonic analogue of the equivalence of (b) and (c) in Theorem 4.6 in the context of harmonic spaces can be found in [GGG2].

7 The Dirichlet Problem with Non-Compact Boundary

7.1 Introduction

Let Ω be an open set in $\mathbf{R}^n$, let $\partial\Omega_{\text{reg}}$ be the set of points in $\partial\Omega$ at which $\mathbf{R}^n\backslash\Omega$ is not thin (cf. Theorem 0.G), and let $\partial\Omega_{\text{irr}} = \partial\Omega\backslash\partial\Omega_{\text{reg}}$. If Ω is bounded, then the Perron-Wiener-Brelot method can be used to solve the Dirichlet problem on Ω, in the sense that for each f in $C(\partial\Omega)$ there is a (unique) h_f in $\mathcal{H}(\Omega)$ such that

$$h_f(X) \to f(Y) \qquad (X \to Y; Y \in \partial\Omega_{\text{reg}}) \tag{7.1}$$

and

$$\limsup_{X\to Y} |h_f(X)| < +\infty \qquad (Y \in \partial\Omega_{\text{irr}}). \tag{7.2}$$

If Ω is unbounded and we consider only boundary functions in the class $C(\partial\Omega\cup\{\infty\})$, where ∞ denotes the Alexandroff point for $\mathbf{R}^n$, then the PWB method can still be used to solve the Dirichlet problem on Ω. However, a more general question concerns whether, for every f in $C(\partial\Omega)$, there exists h_f in $\mathcal{H}(\Omega)$ which satisfies (7.1) and (7.2). The earliest work on this problem seems to have been by R. Nevanlinna [Nev] in 1925. He showed that, for every f in $C(\mathbf{R})$ there is a harmonic function h_f in the upper half-plane satisfying (7.1) (there are no irregular boundary points in this case) and he provided modified Poisson kernels for the construction of h_f. This chapter uses harmonic approximation results from Chapter 3 to give a complete characterization of those open sets Ω in which such a Dirichlet problem can always be solved. We will also discuss the uniqueness or otherwise of the solution h_f when it exists, and a related maximum principle for subharmonic functions on unbounded open sets.

7.2 The Dirichlet Problem

The main result of this chapter is as follows.

Theorem 7.1. *Let Ω be an open set in $\mathbf{R}^n$. The following are equivalent:*
(a) for every f in $C(\partial\Omega)$ there is a harmonic function h_f on Ω which satisfies (7.1) and (7.2);
(b) for each compact set K in $\mathbf{R}^n$ there is a compact set L which contains the bounded components of $\Omega \backslash K$ whose closure intersects K.

In fact, a trivial modification of the proof of Theorem 7.1 below yields a little more: if (b) holds, then for any locally bounded Borel measurable function f on $\partial\Omega$ there exists h_f in $\mathcal{H}(\Omega)$ such that $h_f(X) \to f(Y)$ as $X \to Y$ for each Y in $\partial\Omega_{\text{reg}}$ at which f is continuous, and h_f is bounded near all other points of $\partial\Omega$. We observe that condition (b) of Theorem 7.1 is equivalent to saying that $(\mathbf{R}^n, \mathbf{R}^n \backslash \Omega)$ satisfies the (K, L)-condition.

Example 7.2. Let $n = 2$. Then condition (b) of Theorem 7.1 fails to hold if Ω is replaced by either

$$\Omega_1 = \bigcup_{m=1}^{\infty} \left[\left(\frac{1}{2m+1}, \frac{1}{2m} \right) \times (0, m) \right] \text{ or } \Omega_2 = \Omega_1 \cup \left[\mathbf{R} \times (-\infty, 1) \right].$$

We note that, if $f \in C(\partial\Omega_1)$ and we solve the Dirichlet problem separately in each component V_m of Ω_1, then the resultant function h_f satisfies (7.1) at each point Y of $\cup_m \partial V_m$, but not necessarily at points Y of $\{0\} \times [0, +\infty)$.

Some additional notation will be helpful. If $0 \le r \le R$, then we write $\partial\Omega[r, R]$ for the set $\{X \in \partial\Omega : r \le |X| \le R\}$. If A is a Borel subset of $\mathbf{R}^n$, then we write χ_A for the characteristic function valued 1 on A and 0 elsewhere in $\mathbf{R}^n \cup \{\infty\}$, and write $I(\omega, A)$ for $H^{\omega}_{\chi_A}$. The proof of Theorem 7.1 is given in the next two sections.

7.3 Proof that (b) Implies (a) in Theorem 7.1

Let Ω be an open set in $\mathbf{R}^n$ such that condition (b) of Theorem 7.1 holds and let $f \in C(\partial\Omega)$. There is no loss of generality in assuming that $O \in \Omega$. Let $r_0 = 0$ and $r_1 = 1$. We inductively define an increasing sequence $(r_j)_{j \ge 1}$ of real numbers and a sequence $(F_j)_{j \ge 1}$ of relatively closed subsets of Ω as follows. Given r_j, where $j \ge 1$, let F_j denote the union of $\overline{B(O, r_j)} \cap \Omega$ with all bounded components of $\Omega \backslash \overline{B(O, r_j)}$ whose closure intersects $\overline{B(O, r_j)}$.

By hypothesis F_j is bounded, so we can choose the number r_{j+1} such that $F_j \subset B(O, r_{j+1} - 1)$.

Next we define a sequence $(a_j)_{j\geq 0}$ of non-negative constants by

$$a_j = \begin{cases} \sup\left\{ |f(Y)| : Y \in \partial\Omega[r_j, r_{j+1}] \right\} & \text{if } \partial\Omega[r_j, r_{j+1}] \neq \emptyset \\ 0 & \text{if } \partial\Omega[r_j, r_{j+1}] = \emptyset. \end{cases}$$

We also define the compact sets

$$F_{j,k} = \left\{ Y \in \partial F_j \cap \Omega : \operatorname{dist}(Y, \partial\Omega) \geq 1/k \right\} \qquad (j, k \in \mathbf{N})$$

and the functions

$$s_{j,k}(X) = \begin{cases} a_{j+1} I\big(\Omega\backslash F_{j,k}, \partial\Omega[r_{j+1}, r_{j+2}]\big)(X) & (X \in \Omega\backslash F_{j,k}) \\ 0 & (\text{elsewhere in } \mathbf{R}^n). \end{cases}$$

Since the irregular Euclidean boundary points of $\Omega\backslash F_{j,k}$ form a polar set (by Theorems 0.G and 0.E), the upper regularization $s_{j,k}^*$ of $s_{j,k}$ is subharmonic on $B(O, r_{j+1})$ (see [Doo, 1.V.5]). We observe that

$$I\big(\Omega\backslash F_{j,k}, \partial\Omega[r_{j+1}, r_{j+2}]\big)(X) \downarrow 0 \qquad (k \to \infty; X \in F_j^\circ).$$

Hence the decreasing sequence $(s_{j,k}^*)_{k\geq 1}$ of subharmonic functions on the ball $B(O, r_{j+1})$ has limit 0 on $\overline{B(O, r_j)}$, except possibly on the set $\partial\Omega_{\text{irr}} \cap \overline{B(O, r_j)}$, where $s_{j,k}^*$ might differ from $s_{j,k}$. However, this latter set is polar, and the function $\lim_{k\to\infty} s_{j,k}^*$ is subharmonic on $B(O, r_{j+1})$, so we actually have $s_{j,k}^*(X) \to 0$ as $k \to \infty$ for all X in $\overline{B(O, r_j)}$. It follows from Dini's Theorem that the convergence to 0 of this decreasing sequence of upper semicontinuous functions is uniform on the compact set $\overline{B(O, r_j)}$, and so we can choose k_j in $\mathbf{N}$ such that

$$s_{j,k_j}(X) \leq 2^{-j} \qquad \big(X \in \Omega \cap B(O, r_j)\big). \tag{7.3}$$

Hence the equation

$$v(X) = (a_0 + a_1) I\big(\Omega, \partial\Omega[r_0, r_2]\big)(X) + \sum_{j=1}^{\infty} s_{j,k_j}(X)$$

defines a harmonic function on the open set Ω' given by

$$\Omega' = \Omega\backslash \left(\bigcup_{j=1}^{\infty} F_{j,k_j} \right).$$

This function v is in the upper PWB class for the Dirichlet problem on Ω' with boundary data given by $|g|\chi_{\partial\Omega_{\text{reg}}}$, where

$$g(Y) = \begin{cases} f(Y) & (Y \in \partial\Omega) \\ 0 & (Y \in \cup_j F_{j,k_j} \text{ or } Y = \infty). \end{cases}$$

It follows that g is integrable with respect to harmonic measure on $\partial\Omega'$, and so we can define $u_f = H_g^{\Omega'}$. Thus

$$u_f(X) = H_{g_j}^{\Omega' \cap B(O,r_j)}(X) \qquad (X \in \Omega' \cap B(O,r_j); j \in \mathbf{N}),$$

where

$$g_j(Y) = \begin{cases} g(Y) & (Y \in \partial\Omega' \cap \overline{B(O,r_j)}) \\ u_f(Y) & (Y \in \Omega' \cap \partial B(O,r_j)). \end{cases}$$

We know from (7.3) that v, and hence u_f, is bounded on $\Omega' \cap \partial B(O,r_j)$. Thus

$$u_f(X) = H_{g_j}^{\Omega' \cap B(O,r_j)}(X) \to f(Y) \quad (X \to Y; Y \in \partial\Omega_{\text{reg}} \cap B(O,r_j)).$$

Since j can be arbitrarily large, we obtain

$$u_f(X) \to f(Y) \qquad (X \to Y; Y \in \partial\Omega_{\text{reg}}). \tag{7.4}$$

Similarly it can be seen that

$$\limsup_{X \to Y} |u_f(X)| < +\infty \qquad (Y \in \partial\Omega_{\text{irr}}). \tag{7.5}$$

We can assume that $\Omega \neq \mathbf{R}^n$, for otherwise condition (a) of Theorem 7.1 trivially holds. We now define $d(X) = \text{dist}(X, \mathbf{R}^n \backslash \Omega)$ for each X in Ω, and

$$U_j = \bigcup_{X \in F_{j,k_j}} B\big(X, d(X)/2\big) \qquad (j \in \mathbf{N}).$$

Since F_{j,k_j} is compact, the open set U_j has only finitely many components $U_{j,1}, U_{j,2}, \ldots, U_{j,l_j}$. For each l in $\{1, 2, \ldots, l_j\}$ we choose $X_{j,l}$ in $U_{j,l} \backslash F_j$. It follows from our construction of F_j and our hypothesis that condition (b) holds, that there is a continuous function $p_{j,l} : [0, +\infty) \to \Omega \backslash F_j$ such that $p_{j,l}(0) = X_{j,l}$ and $p_{j,l}(t) \to \infty$ as $t \to +\infty$. (See the remarks following the proof of Lemma 3.2.) We now define

$$V_j = \bigcup_{l=1}^{l_j} \left(\bigcup_{t \geq 0} B\Big(p_{j,l}(t), d\big(p_{j,l}(t)\big)/2\Big) \right) \qquad (j \in \mathbf{N})$$

and $W = \cup_j (U_j \cup V_j)$. Clearly $W \cup \{\infty\}$ is connected and locally connected, and $\overline{W} \subset \Omega$.

For the remainder of this section we assume that $n \geq 3$; the modifications required when $n = 2$ are routine. For each unbounded component ω_m of Ω

we choose Q_m in ω_m. If there are infinitely many unbounded components, then we choose the points Q_m in such a way that $Q_m \to \infty$. Next, for each m, we choose a positive number R_m such that $\overline{B(Q_m, R_m)} \subset \omega_m$, and we define $b_m = \min\{1, 2^{-m} R_m^{n-2}\}$. The function w_1 defined by

$$w_1(X) = \sum_m b_m |X - Q_m|^{2-n} \qquad (X \in \mathbf{R}^n)$$

is a potential on $\mathbf{R}^n$ which is bounded and continuous on $\partial\Omega$. If we define $w_1(\infty) = 0$ and $w_2 = w_1 - H_{w_1}^{\Omega}$ on Ω, then w_2/b_m is equal in ω_m to the Green function for ω_m with pole at Q_m, and w_2 continuously vanishes on $\partial\Omega_{\text{reg}}$. We now apply Corollary 3.6 separately to each of the components ω_m, with Q replaced by Q_m, the set E replaced by $\omega_m \backslash W$, with u replaced by u_f, and with ϵ replaced by b_m. It follows that there is a harmonic function h_0 on $\cup_m \omega_m$ such that

$$\left|h_0(X) - u_f(X)\right| < \min\{w_2(X), 1\} \qquad (X \in (\cup_m \omega_m) \backslash W). \tag{7.6}$$

Finally we define

$$h_f(X) = \begin{cases} h_0(X) & (X \in \cup_m \omega_m) \\ u_f(X) & (X \in \Omega \backslash (\cup_m \omega_m)), \end{cases}$$

and observe that (7.1) and (7.2) follow from (7.4)–(7.6). Thus condition (a) of Theorem 7.1 is established.

7.4 Proof that (a) Implies (b) in Theorem 7.1

Conversely, suppose that condition (b) of Theorem 7.1 fails to hold. Then there is a compact subset K of $\mathbf{R}^n$, and a sequence (U_j) of distinct bounded components of $\Omega \backslash K$ whose closures intersect K, such that $U_j \backslash \overline{B(O, j)} \neq \emptyset$. For each j we choose X_j in U_j such that $\text{dist}(X_j, K) < 1/j$. In fact, without loss of generality, we may assume that $O \in K$ and that $|X_j| < 1/j$ for each j.

Let $A(r, R)$ denote $\{X \in \mathbf{R}^n : r < |X| < R\}$ whenever $0 \leq r < R$, and let $D_j = \partial U_j \cap A(j, j+1)$. Then D_j must have positive harmonic measure for U_j. For, if this were not so, then there would be a superharmonic function v_1 on U_j with limit $+\infty$ at each point of D_j, and the function v_2 defined by

$$v_2(X) = \begin{cases} v_1(X) & (X \in A(j, j+1) \cap U_j) \\ +\infty & (X \in A(j, j+1) \backslash U_j) \end{cases}$$

would be lower semicontinuous and super-meanvalued on $A(j, j+1)$. This is impossible since $A(j, j+1) \cap U_{j+1}$ is a non-polar subset of $A(j, j+1) \backslash U_j$.

Now let

$$u_j = H_{f_j}^{U_j},$$

where

$$f_j(Y) = \begin{cases} |Y| - j & (Y \in A(j, j+1/2)) \\ j+1-|Y| & (Y \in \overline{A(j+1/2, j+1)}) \\ 0 & (\text{elsewhere in } \mathbf{R}^n). \end{cases}$$

It follows from the previous paragraph that $u_j > 0$ on U_j. If we define

$$f(Y) = \sum_{j=1}^{\infty} j f_j(Y)/u_j(X_j) \qquad (Y \in \partial\Omega),$$

then $f \in C(\partial\Omega)$. Suppose now that condition (a) of Theorem 7.1 holds. Then there exists h_f in $\mathcal{H}(\Omega)$ such that (7.1) and (7.2) hold. We define $b = 0$ if $K \cap \Omega = \emptyset$. If $K \cap \Omega \neq \emptyset$, then we choose b to be a negative number such that

$$b \leq \inf\{h_f(Y) : Y \in K \cap \Omega\}.$$

(This is possible because of (7.2).) It is clear from (7.1), (7.2) and the minimum principle applied to U_j that

$$h_f(X) > j u_j(X)/u_j(X_j) + b \qquad (X \in U_j),$$

and so $h_f(X_j) > j + b \to +\infty$ as $j \to \infty$. This contradicts (7.1) and (7.2). Thus condition (a) cannot hold, and the proof of Theorem 7.1 is complete.

7.5 A Maximum Principle

Let s be a subharmonic function on an open set Ω in $\mathbf{R}^n$, and define

$$\overline{s}(Y) = \limsup_{X \to Y} s(X) \qquad (Y \in \partial\Omega).$$

If Ω is bounded, then it follows from the maximum principle that

$$\sup_{\Omega} s = \sup_{\partial\Omega} \overline{s}. \tag{7.7}$$

This equation breaks down for some unbounded sets Ω, as can be seen by considering the function $s(X) = x_n$ on $\mathbf{R}^{n-1} \times (0, +\infty)$, but not all such Ω. The first theorem below characterizes those sets Ω for which (7.7) always holds. The second result, which is closely related, characterizes those open sets Ω for which any solution to the Dirichlet problem in the sense of (7.1) and (7.2) fails to be unique. We will say that ∞ is *accessible from* Ω if there is a continuous function $p : [0, +\infty) \to \Omega$ such that $p(t) \to \infty$ as $t \to +\infty$.

Theorem 7.3. *Let Ω be an open set in $\mathbf{R}^n$. The following are equivalent:*
(a) $\sup_{\Omega} s = \sup_{\partial\Omega} \overline{s}$ for each subharmonic function s on Ω;
(b) ∞ is not accessible from Ω.

Theorem 7.4. *Let Ω be an open set in $\mathbf{R}^n$. The following are equivalent:*
(a) there is a non-constant harmonic function h_0 on Ω such that

$$h_0(X) \to 0 \qquad (X \to Y; Y \in \partial\Omega_{\mathrm{reg}}) \tag{7.8}$$

and

$$\limsup_{X \to Y} |h_0(X)| < +\infty \qquad (Y \in \partial\Omega_{\mathrm{irr}}); \tag{7.9}$$

(b) ∞ is accessible from Ω.

Example 7.5. Let $n = 2$, let Ω_1 be as in Example 7.2 and let $\Omega = \Omega_1 \cup (0,1)^2$. Then ∞ is not accessible from Ω.

Corollary 7.6. *Let Ω be an unbounded connected open set in $\mathbf{R}^n$. If, for each f in $C(\partial\Omega)$, there exists h_f in $\mathcal{H}(\Omega)$ such that (7.1) and (7.2) hold, then for each f in $C(\partial\Omega)$ there are infinitely many distinct members h_f of $\mathcal{H}(\Omega)$ such that (7.1) and (7.2) hold.*

Proof of Corollary. Let Ω be an unbounded connected open set in $\mathbf{R}^n$ such that, for each f in $C(\partial\Omega)$, there exists h_f in $\mathcal{H}(\Omega)$ satisfying (7.1) and (7.2). It follows from Theorem 7.1 that $\Omega \cup \{\infty\}$ is locally connected. Thus (see the remarks following Lemma 3.2) ∞ is accessible from Ω. The result now follows from Theorem 7.4, since $h_f + ch_0$ satisfies (7.1) and (7.2) for any real number c.

Proof of Theorems 7.3 and 7.4. We will prove these two theorems together in three steps by showing that:
(I) if ∞ is accessible from Ω, then condition (a) of Theorem 7.4 holds;
(II) if condition (a) of Theorem 7.4 holds, then condition (a) of Theorem 7.3 fails; and
(III) if ∞ is not accessible from Ω, then condition (a) of Theorem 7.3 holds.

We begin with step (I). Let Ω be an open set in $\mathbf{R}^n$ such that ∞ is accessible from Ω. We can suppose that $\mathbf{R}^n \backslash \Omega$ is non-polar, for otherwise $\partial\Omega_{\mathrm{reg}} = \emptyset$ and condition (a) of Theorem 7.4 trivially holds. In particular, it can be assumed that Ω is Greenian even when $n = 2$. Let $p : [0, +\infty) \to \Omega$ be a continuous function such that $p(t) \to \infty$ as $t \to +\infty$, and let

$$U = \bigcup_{t \geq 0} B\Big(p(t), d(p(t))/2\Big),$$

where $d(X) = \text{dist}(X, \mathbf{R}^n \backslash \Omega)$. We note that $U \cup \{\infty\}$ is connected and locally connected, and that $\overline{U} \subset \Omega$. Next, let ω be the component of Ω which contains U, and let $Q = p(0)$. We apply Corollary 3.6 with $E = (\omega \backslash U) \cup \{Q\}$, with $u = \chi_{B(Q, d(Q)/4)}$ and with $\epsilon = 1/2$, to obtain v in $\mathcal{H}(\omega)$ such that

$$\left|(v-u)(X)\right| < 2^{-1} \min\{1, G_\omega(Q,X)\} \qquad (X \in (\omega \backslash U) \cup \{Q\}),$$

where G_ω is the Green function for ω. If we define h_0 to be equal to v in ω and equal to 0 in $\Omega \backslash \omega$, then $h_0 \in \mathcal{H}(\Omega)$ and (7.8) and (7.9) clearly hold. Further, $h_0(Q) > 1/2$, and h_0 continuously vanishes on the non-empty set $\partial\Omega_{\text{reg}}$, so h_0 is a non-constant function. Thus condition (a) of Theorem 7.4 is established.

Next we turn to step (II). Let h_0 be as in condition (a) of Theorem 7.4 and let Q and r be such that $\overline{B(Q,r)} \subset \Omega$. The set $\partial\Omega_{\text{irr}}$ is polar, so there is a positive superharmonic function v on the Greenian open set $\mathbf{R}^n \backslash \overline{B(Q,r)}$ which takes the value $+\infty$ on $\partial\Omega_{\text{irr}}$ and which continuously vanishes on $\partial B(Q,r)$. Let

$$a = \sup\left\{ \left|h_0(X)\right| : X \in \overline{B(Q,r)} \right\}.$$

For each positive number ϵ the function s_ϵ defined by

$$s_\epsilon(X) = \begin{cases} \left(\left|h_0(X)\right| - a - \epsilon v(X)\right)^+ & (X \in \Omega \backslash \overline{B(Q,r)}) \\ 0 & (X \in \overline{B(Q,r)}) \end{cases}$$

is subharmonic on Ω, and $\overline{s_\epsilon} = 0$ on $\partial\Omega$. If condition (a) of Theorem 7.3 holds, then $s_\epsilon \equiv 0$ on Ω, whence $|h_0| \le a + \epsilon v$ on Ω. Since this holds for any positive number ϵ, it follows that $|h_0| \le a$ on the set $\Omega \backslash \{X : v(X) = +\infty\}$, and hence on Ω. Thus, by the maximum principle, h_0 is constant, which contradicts our hypotheses. We conclude that condition (a) of Theorem 7.3 must fail.

Finally, we come to step (III). Suppose that ∞ is not accessible from Ω, let s be a subharmonic function on Ω, let $M = \sup_{\partial\Omega} \overline{s}$ and let $\epsilon > 0$. We may suppose that $M < +\infty$, for otherwise it is clearly the case that (7.7) holds. We fix m in $\mathbf{N}$, let $W_0 = \Omega \backslash \overline{B(O,m)}$, let $U_{0,1}, U_{0,2}, \ldots$ denote those components of W_0 for which either $\Omega \cap \partial U_{0,k} = \emptyset$ or $s \le M + \epsilon$ on $\Omega \cap \partial U_{0,k}$, and let $V_{0,1}, V_{0,2}, \ldots$ denote the remaining components of W_0. Thus, for each $V_{0,k}$, there exists a point Y_k in $\Omega \cap \partial V_{0,k}$ such that $s(Y_k) > M + \epsilon$. Clearly $Y_k \in \partial B(O,m)$ for each k. There can only be finitely many, k_0 say, of the components $V_{0,k}$, for otherwise there is a subsequence of (Y_k) which converges to some point Y_0 of $\partial\Omega$ and $\overline{s}(Y_0) \ge M + \epsilon$: this is a contradiction.

If $k_0 \ge 1$, then we proceed to define

$$W_1 = \left(\bigcup_{k=1}^{k_0} V_{0,k} \right) \backslash \overline{B(O, m+1)}$$

and divide the components of W_1 as before into two classes $\{U_{1,1}, U_{1,2}, \ldots\}$ and $\{V_{1,1}, V_{1,2}, \ldots, V_{1,k_1}\}$. Similarly, if $k_1 \geq 1$, then we define

$$W_2 = \left(\bigcup_{k=1}^{k_1} V_{1,k} \right) \backslash \overline{B(O, m+2)},$$

and so on. If $j \geq 1$, then each $V_{j,k}$ is a component of W_j and so must be contained in some $V_{j-1,k'}$. Thus, if $j > j'$, each $V_{j,k}$ is contained in some $V_{j',k'}$: in this case we say that $V_{j,k}$ is a *descendent* of $V_{j',k'}$.

Now suppose that, for each j, the collection $\{V_{j,1}, V_{j,2}, \ldots, V_{j,k_j}\}$ is non-empty. It follows that, for some choice of k, the set $V_{0,k}$ has infinitely many descendents: we call this set V_0. There must be a descendent $V_{1,k}$ of V_0 which also has infinitely many descendents: we call this set V_1. Proceeding in this manner, we obtain a sequence $(V_j)_{j \geq 0}$ of connected open subsets of Ω such that $V_0 \supset V_1 \supset \cdots$ and $V_j \cap \overline{B(O, m+j)} = \emptyset$. However, this enables us to construct a continuous function $p : [0, +\infty) \to \Omega$ such that $p(t) \to \infty$ as $t \to +\infty$, and so yields a contradiction.

Thus there exists j' for which there are no sets $V_{j',k}$ as above, in which case $W_{j'} = \cup_k U_{j',k}$ and we do not construct $W_{j'+1}$. If we define

$$K_m = \overline{B(O, m+j')} \backslash \left(\bigcup_{j=0}^{j'-1} \left[\bigcup_k U_{j,k} \right] \right),$$

then K_m is compact, $\overline{B(O, m)} \subset K_m$ and

$$\Omega \backslash K_m = \bigcup_{j=0}^{j'} \left[\bigcup_k U_{j,k} \right].$$

From the definition of the sets $U_{j,k}$, we see that $s \leq M + \epsilon$ on $\Omega \cap \partial K_m$. Hence, by the maximum principle, $s \leq M + \epsilon$ on $\Omega \cap K_m^\circ$ and hence on $\Omega \cap B(O, m)$.

The above argument is valid for each choice of m, so $s \leq M + \epsilon$ on Ω. Since ϵ was arbitrary, $s \leq M$ on Ω, and condition (a) of Theorem 7.3 has been established.

Notes

As was mentioned in §7.1, R. Nevanlinna [Nev] showed that, for every f in $C(\mathbf{R})$, there is a harmonic function h_f on the upper halfplane satisfying (7.1), and he supplied modified Poisson kernels for the construction of h_f (see also [FS]). Analogous results for the half-space can be found in [Gar1]. Theorems 7.1 and 7.4 come from [Gar4]. The implication "(b)⇒(a)" of Theorem 7.3 can be regarded as a corollary of a result of Fuglede [Fug2] on asymptotic paths for subharmonic functions, which was proved using fine potential theory. However, the elementary argument presented above (step (III) in §7.5) is due to Chen and Gauthier [CG]. The converse implication was first observed by Gauthier, Grothmann and Hengartner [GGH].

8 Further Applications

8.1 Non-uniqueness for the Radon Transform

Let f be a real- or complex-valued function on $\mathbf{R}^n$ such that f is integrable on each $(n-1)$-dimensional hyperplane P of $\mathbf{R}^n$. The Radon transform $\widehat{f}$ of f is defined on the collection $\mathcal{P}^{(n)}$ of all such hyperplanes by

$$\widehat{f}(P) = \int_P f\, d\Lambda \qquad (P \in \mathcal{P}^{(n)}),$$

where Λ denotes $(n-1)$-dimensional Lebesgue measure on P. (An account of the Radon transform and its applications can be found in Helgason [Helg].) A long-standing question concerning the Radon transform was whether there exists a non-constant continuous function f such that $\widehat{f} \equiv 0$ on $\mathcal{P}^{(n)}$. The following result asserts that there even exists a non-constant harmonic function with this property.

Theorem 8.1. *There exists a non-constant harmonic function h on $\mathbf{R}^n$ such that $\widehat{h} \equiv 0$ on $\mathcal{P}^{(n)}$.*

We will require the following simple lemma. If $Y' = (y_1, \ldots, y_{n-1})$, then we write dY' for $dy_1 \ldots dy_{n-1}$.

Lemma 8.2. *Let h be a harmonic function on $\mathbf{R}^n$ such that*

$$t \mapsto \int_{\mathbf{R}^{n-1}} |h(Y', t)|\, dY' \qquad (t \in \mathbf{R})$$

defines a locally bounded function on $\mathbf{R}$. Then the equation

$$f(t) = \int_{\mathbf{R}^{n-1}} h(Y', t)\, dY' \qquad (t \in \mathbf{R})$$

defines a polynomial f of degree at most 1.

Proof of Lemma. For each m in $\mathbf{N}$ we define the function

$$h_m(X',x_n) = \int_{\{Y' \in \mathbf{R}^{n-1}:|Y'| \leq m\}} h(X'+Y',x_n)\,dY'$$
$$\big((X',x_n) \in \mathbf{R}^{n-1} \times \mathbf{R}\big).$$

Clearly h_m is continuous and (by Fubini's Theorem) has the mean value property, so $h_m \in \mathcal{H}(\mathbf{R}^n)$. Since

$$\big|h_m(X',x_n)\big| \leq \int_{\mathbf{R}^{n-1}} \big|h(Y',x_n)\big|\,dY' \quad (m \in \mathbf{N};\, (X',x_n) \in \mathbf{R}^{n-1} \times \mathbf{R}),$$

it follows from our hypothesis that the sequence (h_m) is locally uniformly bounded on $\mathbf{R}^n$. Since $h_m(X',x_n) \to f(x_n)$ as $m \to \infty$, the function $(X',x_n) \mapsto f(x_n)$ is harmonic on $\mathbf{R}^n$, whence f is a polynomial of degree at most 1.

Proof of Theorem 8.1. If $Y \in \partial B(O,1)$, then we define the $(n-1)$-dimensional hyperplane

$$P(Y,a) = \{X : \langle X,Y \rangle = a\} \qquad (a \in \mathbf{R})$$

and the strip (or half-space, or whole space)

$$S(Y;a,b) = \{X : \langle X,Y \rangle \in (a,b)\} \qquad (-\infty \leq a < b \leq +\infty).$$

We also define the open set

$$W = \bigcup_{t \geq 0} B\big((t,t^2,\ldots,t^n),1\big),$$

and observe that $W \cup \{\infty\}$ is connected and locally connected. If $Y \in \partial B(O,1)$, where $Y = (y_1,y_2,\ldots,y_n)$, then the function

$$g(t) = \sum_{k=1}^{n} y_k t^k \qquad (t \geq 0)$$

satisfies either $g(t) \to +\infty$ or $g(t) \to -\infty$ as $t \to +\infty$, and it follows that $W \cap S(Y;-a,a)$ is a bounded set for any choice of positive number a. Also, there must exist t_Y such that either $W \cap S(Y;-\infty,t_Y) = \emptyset$ or $W \cap S(Y;t_Y,+\infty) = \emptyset$.

We now define $E = (\mathbf{R}^n \backslash W) \cup \{O\}$ and $u = \chi_{B(O,1/2)}$, so that $u \in \mathcal{H}(E)$, and $(\mathbf{R}^n)^* \backslash E$ is connected and locally connected. It follows from Corollary 5.10 that there exists h in $\mathcal{H}(\mathbf{R}^n)$ such that

$$\big|(h-u)(X)\big| < \big(1+|X|\big)^{-n-1} \qquad (X \in E). \tag{8.1}$$

In particular, $h(O) > 0$ and $h(X) \to 0$ as $X \to \infty$ along E, so h is non-constant. We fix Y in $\partial B(O,1)$ and let $a > 0$. Since $W \cap S(Y;-a,a)$ is bounded, it is contained in $B(O,r)$ for a suitable choice of r. If $|t| < a$, then

$$\begin{aligned}\int_{P(Y,t)} |h|\, d\Lambda &\le \sup\{|h(X)| : |X| \le r\} \int_{P(Y,t)\cap B(O,r)} d\Lambda \\ &\quad + \int_{P(Y,t)\setminus W} \frac{d\Lambda(X)}{\left(1+|X|\right)^{n+1}} \\ &\le \lambda_{n-1} r^{n-1} \sup\left\{|h(X)| : |X| \le r\right\} + \int_{P(Y,0)} \frac{d\Lambda(X)}{\left(1+|X|\right)^{n+1}},\end{aligned}$$

where λ_n denotes the volume of $B(O,1)$. It follows that the function

$$t \mapsto \int_{P(Y,t)} |h|\, d\Lambda \qquad (t \in \mathbf{R})$$

is locally bounded. In particular, h is integrable on $P(Y,t)$ for each t. Further, using Lemma 8.2 and a suitable rotation, we see that $\widehat{h}\big(P(Y,t)\big)$ is a polynomial in t of degree at most 1. Finally, if $P(Y,t) \cap W = \emptyset$, then we use (8.1) to obtain

$$\begin{aligned}\left|\widehat{h}\big(P(Y,t)\big)\right| &< \int_{P(Y,t)} \left(1+|X|\right)^{-n-1} d\Lambda(X) \\ &= \int_{P(Y,0)} \left[1 + \left(|X|^2 + t^2\right)^{1/2}\right]^{-n-1} d\Lambda(X),\end{aligned}$$

and the latter integral tends to 0 as $|t| \to +\infty$. Since there exists t_Y such that either $W \cap S(Y;-\infty,t_Y) = \emptyset$ or $W \cap S(Y;t_Y,+\infty) = \emptyset$, we conclude that $\widehat{h}\big(P(Y,t)\big) \equiv 0$ as a function of t. This holds for all Y in $\partial B(O,1)$, so the theorem is proved.

We remark that, in the course of above proof, we established the following.

Example 8.3. There is a non-constant harmonic function h on $\mathbf{R}^n$ which is bounded on every strip. In fact, for each positive number α, we can choose h_α in $\mathcal{H}(\mathbf{R}^n)$ such that

$$|X|^\alpha \left|h_\alpha(X)\right| \to 0 \qquad \big(|X| \to \infty; X \in S(Y;a,b)\big)$$

for every a, b in $\mathbf{R}$ and every Y in $\partial B(O,1)$.

8.2 A Universal Harmonic Function

This application is elementary, but sufficiently interesting to merit inclusion. Let $Q_k = (k, 0, \ldots, 0)$ in $\mathbf{R}^n$ for each k in $\mathbf{N}$. The following theorem asserts that there is a "universal" harmonic function on $\mathbf{R}^n$; that is, a function h in $\mathcal{H}(\mathbf{R}^n)$ whose translates $X \mapsto h(X + Q_k)$ $(k \in \mathbf{N})$ are dense in $\mathcal{H}(\mathbf{R}^n)$ in the topology of local uniform convergence. The corresponding result for entire holomorphic functions is due to Birkhoff [Bir].

Theorem 8.4. *There is a harmonic function h on $\mathbf{R}^n$ with the following property: for each u in $\mathcal{H}(\mathbf{R}^n)$, each compact set K and each positive number ϵ, there exists k in $\mathbf{N}$ such that*

$$\left|h(X - Q_k) - u(X)\right| < \epsilon \qquad (X \in K).$$

Proof. Let $\{p_{k,m} : m = 1, \ldots, m_k\}$ be a basis for $\mathcal{H}_k$ (the space of all homogeneous harmonic polynomials of degree k) and $\{q_m : m \in \mathbf{N}\}$ be the set of all finite linear combinations of elements of $\cup_k \{p_{k,m} : m = 1, \ldots, m_k\}$ with rational coefficients. Also, let $P_k = (2^k, 0, \ldots, 0)$ in $\mathbf{R}^n$. If $u \in \mathcal{H}(\mathbf{R}^n)$, then we can write u as $\sum_k H_k$, where $H_k \in \mathcal{H}_k$ for each k, and this series converges uniformly on any compact set. By first truncating this series and then approximating the truncated series by a member of $\{q_m : m \in \mathbf{N}\}$ we obtain $q_{m(l)}$ such that

$$\left|q_{m(l)}(X) - u(X)\right| < 2^{-l} \qquad \left(|X| < 2^l\right). \tag{8.2}$$

Indeed, there are infinitely many possible choices of $q_{m(l)}$ for which (8.2) holds. It thus suffices to show that there exists h in $\mathcal{H}(\mathbf{R}^n)$ such that

$$\left|h(X) - q_l(X - P_{l+3})\right| < 2^{-l} \qquad (X \in B(P_{l+3}, 2^l); l \in \mathbf{N}). \tag{8.3}$$

This is immediate from either Theorem 4.6 or Theorem 4.8, but it is unnecessary to use such powerful results.

Instead we proceed as follows. Let

$$B_l = \overline{B(O, 3.2^{l+1})},\ C_l = \overline{B(P_{l+3}, 2^l)},\ \text{and } K_l = B_l \cup C_l \quad (l \in \mathbf{N}).$$

We note that $B_l \cap C_l = \emptyset$ and that $C_l \subset B_{l+1}$. If we define $u_1(X) = q_1(X - P_4)$ on a neighbourhood of C_1 and $u_1 = 0$ on a neighbourhood of B_1, then $u_1 \in \mathcal{H}(K_1)$ so we can use Theorem 1.7 to obtain v_1 in $\mathcal{H}(\mathbf{R}^n)$ such that

$$\left|v_1(X) - u_1(X)\right| < 2^{-2} \qquad (X \in K_1).$$

We now proceed inductively. Given $v_1, v_2, \ldots, v_{l-1}$, we define

$$u_l(X) = \begin{cases} q_l(X - P_{l+3}) - v_1(X) - \cdots - v_{l-1}(X) & (X \in C_l) \\ 0 & (X \in B_l) \end{cases} \tag{8.4}$$

and use Theorem 1.7 to obtain v_l in $\mathcal{H}(\mathbf{R}^n)$ such that

$$|v_l(X) - u_l(X)| < 2^{-l-1} \qquad (X \in K_l). \tag{8.5}$$

Defining

$$h(X) = \sum_{l=1}^{\infty} v_l(X) \qquad (X \in \mathbf{R}^n),$$

we note from (8.4) and (8.5) that $|v_l| < 2^{-l-1}$ on B_l, and so this series converges locally uniformly on $\mathbf{R}^n$. Hence $h \in \mathcal{H}(\mathbf{R}^n)$. If $X \in C_l$, then $X \in B_k$ whenever $k > l$, so

$$\begin{aligned} |h(X) - q_l(X - P_{l+3})| &\leq \left| \sum_{k=1}^{l} v_k(X) - q_l(X - P_{l+3}) \right| + \sum_{k=l+1}^{\infty} |v_k(X)| \\ &< |v_l(X) - u_l(X)| + 2^{-l-1} \\ &< 2^{-l}, \end{aligned}$$

using (8.4) and (8.5). Hence (8.3) holds, and the theorem is proved.

8.3 Boundary Cluster Sets of Subharmonic Functions

Let Ω be an open set in $\mathbf{R}^n$ and $u : \Omega \to [-\infty, +\infty]$. If $A \subseteq \Omega$ and $Z \in \overline{A} \cap \partial\Omega$, then the *cluster set of u along A at Z* is defined by

$$\begin{aligned} C_A(u, Z) = \{ l \in [-\infty, +\infty] : u(X_m) \to l \text{ for some} \\ \text{sequence } (X_m) \text{ of points in } A \text{ such that } X_m \to Z \}. \end{aligned}$$

Particular interest attaches to sets A with the property that $C_A(u, Z) = C_\Omega(u, Z)$. Let $\mathbf{D} = \{z \in \mathbf{C} : |z| < 1\}$. We recall that a subset of a metric space is called *residual* if it is the complement of a first category set. A classical result of Collingwood [CL, p.76] is as follows.

Collingwood Maximality Theorem. *If $u : \mathbf{D} \to \mathbf{R}$ is continuous and if $\{\gamma_\theta : \theta \in [0, 2\pi)\}$ denotes the collection of rotations of an arc γ_0 in $\mathbf{D}$ with one endpoint at 1, then $C_{\gamma_\theta}(u, e^{i\theta}) = C_{\mathbf{D}}(u, e^{i\theta})$ for a residual set of values θ in $[0, 2\pi)$.*

This result fails if u is assumed merely to be upper (or lower) semicontinuous (see [CL, p. 78]), but it is valid if "continuous" is replaced by "subharmonic", (this was first observed by Arsove [Ars]). In higher dimensions arcs must be replaced by "thicker" sets if we are to obtain a corresponding result for subharmonic functions. Just how substantial these sets must be is identified precisely below. It will be notationally more convenient to work in the context of functions defined on the half-space $W = \{(X', x_n) : X' \in \mathbf{R}^{n-1}, x_n > 0\}$.

A subset A_O of W will be called *sectionally non-polar near the boundary* if there is a positive number c such that $A_O \cap \{(X', x_n); a \le x_n \le b\}$ is a non-polar set whenever $0 < a < b < c$. We also define translations of a set A_O by $A_Z = \{X : X - Z \in A_O\}$ for each Z in ∂W.

Theorem 8.5. *Let A_O be a bounded, relatively closed subset of W such that $\overline{A_O} \cap \partial W = \{O\}$. The following are equivalent:*
(a) for every subharmonic function u on W, the equation $C_{A_Z}(u, Z) = C_W(u, Z)$ holds for a residual set of points Z in ∂W;
(b) for every finely continuous function u on W, the equation $C_{A_Z}(u, Z) = C_W(u, Z)$ holds for a residual set of points Z in ∂W;
(c) A_O is sectionally non-polar near the boundary.

We will require the following lemma.

Lemma 8.6. *Let A_O be a subset of W which is sectionally non-polar near the boundary and which satisfies $\overline{A_O} \cap \partial W = \{O\}$. Further, let $r > 0$, let F be a dense subset of $B(O, r) \cap \partial W$, let $q > 0$ and let*

$$A = \bigcup_{Z \in F} \{(X', x_n) \in A_Z : 0 < x_n < q\}.$$

Then there is a positive number t_0 (depending on A_O, r and q) such that the fine closure of A contains the cylinder $\{(X', x_n) : |X'| < r/2, 0 < x_n < t_0\}$.

Proof of Lemma. There is a positive constant c such that the set

$$A_O \cap \{(X', x_n) : a < x_n < b\}$$

is non-polar whenever $0 < a < b < c$. It follows from Theorem 0.E that there exists a dense subset T of $(0, c)$ such that, for each t in T, there is a point (Y', t) in A_O at which A_O is non-thin. We now choose t_0 such that $0 < t_0 < \min\{q, c\}$ and such that

$$A_O \cap \{(X', x_n) : 0 < x_n < t_0\} \subset K,$$

where K is the cylinder in the statement of the lemma.

Let $X_0 = (X_0', t) \in K$, where $t \in T$. There exists (Y_t', t) in A_O at which A_O is non-thin. Since $|X_0' - Y_t'| \le |X_0'| + |Y_t'| < r$, there is a sequence

$((Z'_m, 0))$ of points in F such that $Z'_m \to X'_0 - Y'_t$, whence $(Z'_m + Y'_t, t) \to X_0$ as $m \to \infty$. If ρ is large, then

$$\left\{X \in A : \phi_n(|X - X_0|) > \rho\right\} \supseteq \left\{X \in A_{(Z'_m,0)} : 0 < x_n < q \right.$$
$$\left. \text{and } \phi_n\Big(\big|X - (Z'_m + Y'_t, t)\big|\Big) > 2\rho\right\}$$

for all sufficiently large m, and so

$$\mathcal{C}^*\left(\left\{X \in A : \phi_n(|X - X_0|) > \rho\right\}\right)$$
$$\geq \mathcal{C}^*\left(\left\{X \in A_O : \phi_n\big|X - (Y'_t, t)\big|\big) > 2\rho\right\}\right)$$

where $\mathcal{C}^*$ denotes outer capacity with respect to $B(O, r + q + 1)$. It follows from Theorem 0.J $\big((\text{i}) \Leftrightarrow (\text{iii})\big)$ that A is non-thin at X_0.

We have now shown that the fine closure of A contains the set

$$S = \{(X', t) \in \mathbf{R}^{n-1} \times T : |X'| < r/2 \text{ and } 0 < t < t_0\}.$$

Since an $(n-1)$-dimensional hyperplane is not thin at any of its constituent points, and since T is dense in $(0, c)$, we can repeat the argument of the preceding paragraph to see that the fine closure of S contains K. The lemma is now established.

Proof of Theorem 8.5. Let A_O be a bounded, relatively closed subset of W such that $\overline{A_O} \cap \partial W = \{O\}$ and suppose that condition (c) holds. Also, let u be finely continuous on W, and let

$$F = \left\{Z \in \partial W : C_{A_Z}(u, Z) \neq C_W(u, Z)\right\}.$$

Let $\mathcal{I}$ denote the family of closed intervals in $[-\infty, +\infty]$ with endpoints in $\mathbf{Q} \cup \{-\infty, +\infty\}$ and let $Z \in F$. Noting that $C_{A_Z}(u, Z)$ is a compact subset of $[-\infty, +\infty]$ we can find I in $\mathcal{I}$, and a finite union J of intervals from $\mathcal{I}$ and a positive rational number q such that $I \cap C_W(u, Z) \neq \emptyset$, such that

$$\left\{u(X', x_n) : (X', x_n) \in A_Z \text{ and } 0 < x_n < q\right\} \subset J, \tag{8.6}$$

and such that $I \cap J = \emptyset$. If I, J, q are as above, then we say that $Z \in F(I, J, q)$. Thus we know that

$$F \subset \bigcup_{I,J,q} F(I, J, q),$$

where the union is over all possible choices of I, J, q as described above. Now suppose that one of these sets $F(I_0, J_0, q_0)$ is dense in $B(Z_0, r) \cap \partial W$ for some Z_0 in ∂W and some positive number r, and define

$$A = \bigcup_{Z \in F(I_0, J_0, q_0)} \{(X', x_n) \in A_Z : 0 < x_n < q_0\}.$$

It follows from Lemma 8.6 that the fine closure of A contains the cylinder

$$\{(X', x_n) : |X' - Z_0'| < r/2 \text{ and } 0 < x_n < t_0\},$$

where $Z_0 = (Z_0', 0)$ and t_0 is some positive number depending on A_O, r and q_0. Hence, by (8.6),

$$C_W(u, Z_0) \subset J \subset [-\infty, +\infty] \backslash I,$$

which contradicts the fact that $I \cap C_W(u, Z_0) \neq \emptyset$. Thus the sets $F(I, J, q)$ must all be nowhere dense in ∂W, whence F is of first category in ∂W, and $\partial W \backslash F$ is residual. This establishes (b).

It follows from the definition of the fine topology that (b) implies (a).

It remains to show that (a) implies (c), and we will establish this by contradiction. Suppose that (c) fails. Thus there are positive sequences (a_m) and (b_m), converging to 0, such that $b_1 > a_1 > b_2 > a_2 > \cdots$ and the set

$$A_O \cap \{(X', x_n) : a_m \leq x_n \leq b_m\}$$

is polar for each m. Let $c_m = (a_m + b_m)/2$ and define

$$E = \bigcup_{m=1}^{\infty} (\mathbf{R}^{n-1} \times [b_{m+1}, a_m])$$

and

$$h(X) = -m \qquad (X \in \mathbf{R}^{n-1} \times (c_{m+1}, c_m); m \in \mathbf{N}).$$

Clearly E is a relatively closed subset of W such that $W^* \backslash E$ is connected and locally connected. Also, $h \in \mathcal{H}(E)$. Thus, by Corollary 3.8, there exists v in $\mathcal{H}(W)$ such that

$$-m - 1 < v(X', x_n) < -m + 1 \qquad (x_n \in [b_{m+1}, a_m]; m \in \mathbf{N}).$$

Next we observe that the set

$$S = \bigcup_{m=1}^{\infty} \left(A_O \cap \{(X', x_n) : a_m \leq x_n \leq b_m\} \right)$$

is polar, so there is a positive superharmonic function s on W which is valued $+\infty$ at each point of S. It can easily be arranged that $s(X', x_n) \le 1$ whenever $x_n \ge b_1 + 1$. Now let $\{Z_k : k \in \mathbf{N}\}$ be a dense subset of ∂W and define

$$u(X) = v(X) - \sum_{k=1}^{\infty} 2^{-k} s(X - Z_k) \qquad (X \in W).$$

We observe that u is subharmonic on W and takes the value $-\infty$ on the set

$$\left\{ (X', x_n) \in A_{Z_k} : a_m \le x_n \le b_m \right\}$$

for each choice of m and k in $\mathbf{N}$.

For each m in $\mathbf{N}$ and each k in $\{m+1, m+2, \ldots\}$ we define

$$F_{m,k} = \left\{ Z \in \partial W : u(X) < 1 - m \text{ when } X \in A_Z \text{ and } a_k \le x_n \le a_m \right\}.$$

Since $u(X) < 1 - m$ when $X \in E$ and $x_n \le a_m$, it is clear that

$$F_{m,k} = \bigcup_{j=m+1}^{k} \left\{ Z \in \partial W : u(X) < 1 - m \text{ when } X \in A_Z \text{ and } a_j \le x_n \le b_j \right\}.$$

Noting that the set $\{X \in A_Z : a_j \le x_n \le b_j\}$ is compact and that the set $\{X \in W : u(X) < 1 - m\}$ is open, we see that each set $F_{m,k}$ is relatively open in ∂W. Also, $\{Z_j : j \in \mathbf{N}\} \subseteq F_{m,k}$ for each m and k, so the G_δ set F' defined by

$$F' = \bigcap_{m=1}^{\infty} \bigcap_{k=m+1}^{\infty} F_{m,k}$$

is residual (and hence of second category) in ∂W.

From the definition of F' it is clear that $C_{A_Z}(u, Z) = \{-\infty\}$ for each Z in F'. Now suppose that there exists Z_0 in F' such that $C_W(u, Z_0) = \{-\infty\}$. Then there exists r_0 such that $u(X) < 0$ on $B(Z_0, r_0) \cap W$. Let

$$\beta(t) = \{(X', x_n) \in B((Z_0', t), r_0/2) : x_n > t\},$$

where $Z_0 = (Z_0', 0)$. If $(X', x_n) \in \beta(0)$ and $x_n > a_m$, then

$$\begin{aligned} u(X', x_n) &\le H_u^{\beta(a_m)}(X', x_n) \\ &\le (1 - m) H_{\chi_{\partial W}}^{\beta(0)}(X', x_n - a_m) \to -\infty \qquad (m \to \infty). \end{aligned}$$

Thus $u \equiv -\infty$ on $\beta(0)$, which yields a contradiction. Hence $C_W(u, Z) \ne \{-\infty\}$ for each Z in F'. It is now clear that condition (a) fails to hold, so Theorem 8.5 is proved.

8.4 Growth of Harmonic Functions along Rays

Suppose that h is a harmonic function on $\mathbf{R}^n$. If F is a second category subset of $\partial B(O,1)$, then h cannot satisfy

$$r^{-a}|h(rY)| \to +\infty \qquad (r \to +\infty; Y \in F; a \in (0,+\infty)). \tag{8.7}$$

To see this, suppose that (8.7) holds, let

$$F^+ = \{Y \in F : r^{-a}h(rY) \to +\infty \text{ as } r \to +\infty \text{ for all positive numbers } a\}$$

and let $F^- = F \backslash F^+$. Then one of the sets F^+, F^- must be of second category in $\partial B(O,1)$. We suppose, without loss of generality, that F^+ has this property and define

$$F_j^+ = \{Y \in \partial B(O,1) : h(rY) \geq 1 \text{ whenever } r \geq j\} \qquad (j \in \mathbf{N}).$$

The sets F_j^+ are closed, and $F^+ \subset \cup_j F_j^+$. Thus there exists j' in $\mathbf{N}$ and an open circular cone C_1 (with O as vertex) such that $\overline{C_1} \cap \partial B(O,1) \subset \overline{F^+ \cap F_{j'}^+}$. The function $|h|$ is bounded, by M say, on $\overline{B(O,j')}$, so we can define a positive harmonic function h_1 on C_1 by $h_1(X) = h(X) + M$. Let $Y_1 \in C_1 \cap \partial B(O,1)$. It follows from Harnack's inequalities that $h_1(rY_1) \leq cr^b$ when $r \geq 1$, for some positive numbers b (depending on Y_1 and C_1) and c. This leads to the desired contradiction, in view of Harnack's inequalities and the definition of F^+.

The sharpness of the above observation is illustrated by the following example.

Example 8.7. For each positive number a there exists h in $\mathcal{H}(\mathbf{R}^n)$ such that

$$r^{-a}h(X + rY) \to +\infty \qquad (r \to +\infty) \tag{8.8}$$

for all X in $\mathbf{R}^n$ and all Y in $\partial B(O,1)$.

Details. We deal first with the case where $n = 2$ and identify $\mathbf{R}^2$ with $\mathbf{C}$ in the usual way. Let $a > 0$, let m be a positive integer such that $m > 2a$, and define

$$U_k = \{re^{i(\theta+k\pi)/m} : r > 0 \text{ and } |\theta| < \pi/2\} \quad (k = 0, 1, \ldots, 2m-1),$$

$$U = \bigcup_{k=0}^{2m-1} U_k \text{ and } F = \{re^{i\theta} : r^m|\cos m\theta| \geq 1\}.$$

Thus F is a closed subset of U. We also define

$$u(re^{i\theta}) = r^{m/2}\cos(m\theta/2) \qquad (|\theta| < \pi/(2m)),$$

$$u(z) = u(ze^{-ik\pi/m}) \qquad (z \in U_k; k = 1, 2, \ldots, 2m-1),$$

and

$$u(z) = |z|^{m/2} \qquad (z \in \mathbf{C}\backslash U).$$

Thus $u \in C(E) \cap \mathcal{H}(E^\circ)$, where $E = F \cup (\mathbf{C}\backslash U)$. Since $\mathbf{C}^*\backslash E$ is connected and locally connected we can apply Corollary 3.21 (see the remark following that result) to obtain h_2 in $\mathcal{H}(\mathbf{C})$ such that $|h_2 - u| < 1$ on E. We note that $u(z) \geq 2^{-1/2}|z|^{m/2}$ on E. Also, if $z_0 \in \mathbf{C}$ and $\theta \in \mathbf{R}$, then $z_0 + re^{i\theta} \in E$ for all sufficiently large r. This establishes Example 8.7 when $n = 2$.

We now proceed by induction on n. Let a and m be as above, suppose that $n \geq 3$ and that h_{n-1} is a harmonic function on $\mathbf{R}^{n-1}$ with the properties described in Example 8.7. Let

$$F_1 = \{X \in \mathbf{R}^n : |x_n| \geq 1 + x_1^2 + \cdots + x_{n-1}^2\},$$

$$F_2 = \{X \in \mathbf{R}^n : \mathrm{dist}(X, F_1) \geq 1\},$$

let $F_3 = F_1 \cup F_2$, and let V_1, V_2 be disjoint open neighbourhoods of F_1, F_2 respectively. We define

$$v(X) = \begin{cases} h_{n-1}(x_1, x_2, \ldots, x_{n-1}) & (X \in V_2) \\ (x_1^2 + x_n^2)^{m/2}\cos\left(m\cos^{-1}\left[x_n(x_1^2 + x_n^2)^{-1/2}\right]\right) & (X \in V_1), \end{cases}$$

and observe that $v \in \mathcal{H}(V_1 \cup V_2)$. Since $(\mathbf{R}^n)^*\backslash F_3$ is connected and locally connected, we can apply Corollary 3.8 to obtain h_n in $\mathcal{H}(\mathbf{R}^n)$ such that $|h_n - v| < 1$ on F_3. Let $X \in \mathbf{R}^n$ and $Y \in \partial B(O, 1)$. If $Y = (0, \ldots, 0, \pm 1)$, then $X + rY \in F_1$ for all sufficiently large r. For other choices of Y, we have $X + rY \in F_2$ for all sufficiently large r. In either case we have $X + rY \in F_3$ and it follows from the properties of h_{n-1} that (8.8) holds.

Notes

Zalcman [Zal] used holomorphic approximation to show that there exists a non-constant entire holomorphic function f whose Radon transform $\widehat{f}$ is 0 on $\mathcal{P}^{(2)}$ (identifying $\mathbf{C}$ with $\mathbf{R}^2$ in the usual way). If we define h to be the real part of f, then we obtain Theorem 8.1 when $n = 2$. The higher dimensional result is due to Armitage and Goldstein [AG2]. In the proof of Theorem 8.1 it was convenient to appeal to Corollary 5.10 to obtain a function h in $\mathcal{H}(\mathbf{R}^n)$ such that (8.1) holds. However, it is not difficult to construct such a function h using only the more elementary Lemma 5.6: see

[Arm2]. It is also known that, if f is continuous and, in addition, integrable on $\mathbf{R}^n$, then $f \equiv 0$ if and only if $\widehat{f} \equiv 0$ on $\mathcal{P}^{(n)}$ (see [Zal] for references). However, if $1 \leq m \leq n-2$, and if f is continuous on $\mathbf{R}^n$ and has integral 0 over m-dimensional affine set, it is not known whether f must be identically valued 0.

Theorem 8.3 and its proof are a routine adaptation from Birkhoff [Bir], who proved the corresponding result for entire holomorphic functions.

Rippon [Rip] showed that (c) implies (b) in Theorem 8.4 (the assumption that A_O is relatively closed is not needed for this step) and gave some applications to uniqueness theorems for subharmonic functions in half-spaces. He also raised the question of whether the hypothesis, that A_O be sectionally non-polar near the boundary, was necessary. This was answered affirmatively in [Gar2, §8], where it was shown that (a) implies (c).

Example 8.7 is due to Armitage and Goldstein [AG4]. The preceding observation can be found in [Sch] and [AG4].

References

[Ara1] N. U. Arakeljan: Uniform and tangential approximations by analytic functions. Izv. Akad. Nauk Armjan. SSR Ser. Mat. 3 (1968), 273–286 (Russian). [Amer. Math. Soc. Transl. (2) 122 (1984), 85–97.]

[Ara2] N. U. Arakeljan: Approximation complexe et propriétés des fonctions analytiques. Actes, Congrès intern. Math., (1970), Tome 2, 595–600.

[Arm1] D. H. Armitage: On the extension of superharmonic functions. J. London Math. Soc. (2) 4 (1971), 215–230.

[Arm2] D. H. Armitage: A non-constant continuous function on the plane whose integral on every line is zero. Amer. Math. Monthly, to appear.

[ABG] D. H. Armitage, T. Bagby and P. M. Gauthier: Note on the decay of solutions of elliptic equations. Bull. London Math. Soc. 17 (1985), 554–556.

[AG1] D. H. Armitage and M. Goldstein: Better than uniform approximation on closed sets by harmonic functions with singularities. Proc. London Math. Soc. (3) 60 (1990), 319–343.

[AG2] D. H. Armitage and M. Goldstein: Nonuniqueness for the Radon transform. Proc. Amer. Math. Soc., 117 (1993), 175–178.

[AG3] D. H. Armitage and M. Goldstein: Tangential harmonic approximation on relatively closed sets. Proc. London Math. Soc., (3) 68 (1994), 112–126.

[AG4] D. H. Armitage and M. Goldstein: Radial limiting behaviour of harmonic functions in cones. Complex Variables, Theory Appl. 22 (1993), 267–276.

[Ars] M. G. Arsove: The Lusin-Privalov theorem for subharmonic functions. Proc. London Math. Soc. (3) 14 (1964), 260–270.

[BB] T. Bagby and P. Blanchet: Uniform harmonic approximation on Riemannian manifolds. J. Analyse Math., 62 (1994), 47–76.

[BG1] T. Bagby and P. M. Gauthier: Approximation by harmonic functions on closed subsets of Riemann surfaces. J. Analyse Math. 51 (1988), 259–284.

[BG2] T. Bagby and P. M. Gauthier: Uniform approximation by global harmonic functions. In Approximation by Solutions of Partial Differential Equations, pp.15–26, ed. B. Fuglede et al., NATO ASI Series, Kluwer, Dordrecht, 1992.

[BG3] T. Bagby and P. M. Gauthier: Tangential harmonic approximation on Riemannian manifolds: necessary conditions. Preprint.

[BeG] C. Bensouda and P. M. Gauthier: Approximation surharmonique sur les fermés. Izv. Akad. Nauk Arm. Mat., to appear.

[Bir] G. D. Birkhoff: Démonstration d'un théorème élémentaire sur les fonctions entières. C. R. Acad. Sci. Paris 189 (1929), 473–475.

[BH] J. Bliedtner and W. Hansen: Simplicial cones in potential theory II (Approximation theorems). Invent. Math. 46 (1978), 255–275.

[Bre1] M. Brelot: Sur l'approximation et la convergence dans la théorie des fonctions harmoniques ou holomorphes. Bull. Soc. Math. France 73 (1945), 55–70.

[Bre2] M. Brelot: *Éléments de la théorie classique du potentiel.* Fourth Edition. Centre de Documentation Universitaire, Paris, 1969.

[CarL] L. Carleson: Mergelyan's theorem on uniform polynomial approximation. Math. Scand. 15 (1964), 167–175.

[CarT] T. Carleman: Sur un théorème de Weierstrass. Ark. Mat. Astronom. Fys. 20B (1927), 1–5.

[CG] Chen Huaihui and P. M. Gauthier: A maximum principle for subharmonic and plurisubharmonic functions. Canad. Math. Bull. 35 (1992), 34–39.

[CL] E. F. Collingwood and A. J. Lohwater: *The theory of cluster sets.* Cambridge University Press, 1966.

[Con] J. B. Conway: *Functions of one complex variable.* Second Edition. Springer, New York, 1978.

[DG] A. Debiard and B. Gaveau: Potentiel fin et algèbres de fonctions analytiques I. J. Funct. Anal. 16 (1974), 289–304.

[De1] J. Deny: Sur l'approximation des fonctions harmoniques. Bull. Soc. Math. France 73 (1945), 71–73.

[De2] J. Deny: Systèmes totaux de fonctions harmoniques. Ann. Inst. Fourier (Grenoble) 1 (1949) 103–113.

[Doo] J. L. Doob: *Classical potential theory and its probabilistic counterpart.* Springer, New York, 1983.

[DuP] N. Du Plessis: *An introduction to potential theory.* Oliver and Boyd, Edinburgh, 1970.

[EdVS] D. A. Edwards and G. Vincent-Smith: A Weierstrass-Stone theorem for Choquet simplexes. Ann. Inst. Fourier (Grenoble) 18, 1 (1968), 261–282.

[EK] E. G. Effros and J. L. Kazdan: Applications of Choquet simplexes to elliptic and parabolic boundary value problems. J. Diff. Equations 8 (1970), 95–134.

[FS] M. Finkelstein and S. Scheinberg: Kernels for solving problems of Dirichlet type in a half-plane. Adv. in Math. 18 (1975), 108–113.

[Fug1] B. Fuglede: *Finely harmonic functions.* Lecture Notes in Mathematics 289, Springer, Berlin, 1972.

[Fug2] B. Fuglede: Asymptotic paths for subharmonic functions. Math. Ann. 213 (1975), 261–274.

[Gai] D. Gaier: *Lectures on complex approximation.* Birkhäuser, Boston, 1987.

[Gar1] S. J. Gardiner: The Dirichlet and Neumann problems for harmonic functions in half-spaces. J. London Math. Soc. (2) 24 (1981), 502–512.

[Gar2] S. J. Gardiner: Uniqueness and extension theorems for subharmonic functions. J. London Math. Soc. (2), 48 (1993), 515–525.

[Gar3] S. J. Gardiner: Superharmonic extension and harmonic approximation. Ann. Inst. Fourier (Grenoble), 44 (1994), 65–91.

[Gar4] S. J. Gardiner: The Dirichlet problem with non-compact boundary. Math. Z., 213 (1993), 163–170.

[Gar5] S. J. Gardiner: Tangential harmonic approximation on relatively closed sets. Illinois J. Math., to appear.

[Gar6] S. J. Gardiner: Uniform and tangential harmonic approximation. In Classical and Modern Potential Theory, pp. 185–198, ed. K. GowriSankaran et al., NATO ASI Series, Kluwer, Dordrecht, 1994.

[GG] S. J. Gardiner and M. Goldstein: Carleman approximation by harmonic functions. Amer. J. Math., to appear.

[GGG1] S. J. Gardiner, M. Goldstein and K. GowriSankaran: Global approximation in harmonic spaces. Proc. Amer. Math. Soc., 122 (1994), 213–221.

[GGG2] S. J. Gardiner, M. Goldstein and K. GowriSankaran: Tangential approximation in harmonic spaces. Indiana Univ. Math. J., to appear.

[Gau1] P. M. Gauthier: Tangential approximation by entire functions and functions holomorphic in a disc. Izv. Akad. Nauk Armjan. SSR Ser. Mat. 4 (1969), 319–326.

[Gau2] P. M. Gauthier: Approximation by (pluri)subharmonic functions: fusion and localization. Can. J. Math. 44 (1992), 941–950.

[Gau3] P. M. Gauthier: Subharmonic extensions and approximations. Canad. Math. Bull., 37 (1994), 46–53.

[GGO1] P. M. Gauthier, M. Goldstein and W. H. Ow: Uniform approximation on unbounded sets by harmonic functions with logarithmic singularities. Trans. Amer. Math. Soc. 261 (1980), 169–183.

[GGO2] P. M. Gauthier, M. Goldstein and W. H. Ow: Uniform approximation on closed sets by harmonic functions with Newtonian singularities. J. London Math. Soc. (2) 28 (1983), 71–82.

[GGH] P. M. Gauthier, R. Grothmann and W. Hengartner: Asymptotic maximum principles for subharmonic and plurisubharmonic functions. Can. J. Math. 40 (1988), 477–486.

[GH] P. M. Gauthier and W. Hengartner: *Approximation uniforme qualitative sur des ensembles non bornées.* Séminaire de Mathématiques Supérieures, Press. Univ. Montreal, 1982.

[GHS] P. M. Gauthier, W. Hengartner and A. Stray: A problem of Rubel concerning approximation on unbounded sets by entire functions. Rocky Mountain J. Math. 19 (1989), 127–135.

[GO] M. Goldstein and W. H. Ow: A characterization of harmonic Arakelyan sets. Proc. Amer. Math. Soc., 119 (1993), 811–816.

[Han] W. Hansen: Harmonic and superharmonic functions on compact sets. Illinois J. Math. 29 (1985), 103–107.

[Hed] L. I. Hedberg: Approximation by harmonic functions, and stability of the Dirichlet problem. Expo. Math. 11 (1993), 193–259.

[Helg] S. Helgason: *The Radon transform.* Birkhaüser, Boston, 1980.

[Helm] L. L. Helms: *Introduction to potential theory.* Krieger, New York, 1975.

[HV] Ho-Van Thi Si: Frontière de Choquet dans les espaces de fonctions et approximation des fonctions harmoniques. Bull. Soc. Roy. Sci. Liège 41 (1972), 607–639.

[Ive] F. Iverson: *Recherches sur les fonctions inverses des fonctions méromorphes.* Thesis, Helsingfors, 1914.

[Kap] W. Kaplan: Approximation by entire functions. Michigan Math. J. 3 (1955/6), 43–52.

[Kel] M. V. Keldyš: On the solvability and stability of the Dirichlet problem. Uspehi Mat. Nauk 8 (1941), 171–231 (Russian). [Amer. Math. Soc. Transl. 51 (1966), 1–73.]

[KM] J. Korevaar and J. L. H. Meyers: Logarithmic convexity for supremum norms of harmonic functions. Bull. London Math. Soc., to appear.

[Lab] M. Labrèche: *De l'approximation harmonique uniforme*. Doctoral thesis, Université de Montréal, 1982.

[Lan] N. S. Landkof: *Foundations of modern potential theory*. Springer, New York, 1972.

[Leb] H. Lebesgue: Sur le problème de Dirichlet. Rend. Circ. Mat. di Palermo 29 (1907), 371–402.

[Lyo] T. Lyons: Finely harmonic functions need not be quasi-analytic. Bull. London Math. Soc., 16 (1984), 413–415.

[Mer] S. N. Mergelyan: Uniform approximations to functions of a complex variable. Uspehi Mat. Nauk (N.S.) 7, no. 2 (48) 31–122 (Russian). [Amer. Math. Soc. Transl. 3 (1962), 294–391.]

[Mul] C. Müller: *Spherical harmonics*. Lecture Notes in Math. 17, Springer, Berlin, 1966.

[Ner1] A. A. Nersesyan: Carleman sets. Izv. Akad. Nauk Armjan. SSR Ser. Mat. 6 (1971), 465–471 (Russian). [Amer. Math. Soc. Transl. (2) 122 (1984), 99–104.]

[Ner2] A. A. Nersesyan: Harmonic approximation and the solution of a problem of L. A. Rubel. Dokl. Akad. Nauk. Arm. SSR 84 (1987), 104–106 (Russian).

[Nev] R. Nevanlinna: Über eine Erweiterung des Poissonschen Integrals. Ann. Acad. Sci. Fenn. Ser. A. 24, No. 4 (1925), 1–15.

[PV] P. V. Paramonov and J. Verdera: Approximation by solutions of elliptic equations on closed subsets of Euclidean space. Preprint.

[Pra1] A. de la Pradelle: Approximation et caractère de quasi-analyticité dans la théorie axiomatique des fonctions harmoniques. Ann. Inst. Fourier (Grenoble) 17, 1 (1967), 383–399.

[Pra2] A. de la Pradelle: A propos du mémoire de G. F. Vincent-Smith sur l'approximation des fonctions harmoniques. Ann. Inst. Fourier (Grenoble) 19, 2 (1969), 355–370.

[Rip] P. J. Rippon: The boundary cluster set of subharmonic functions. J. London Math. Soc. (2) 17 (1978), 469–479.

[Rot1] A. Roth: Approximationseigenschaften und Strahlengrenzwerte meromorpher und ganzer Funktionen. Comment. Math. Helv. 11 (1938), 77–125.

[Rot2] A. Roth: Meromorphe Approximationen. Comment. Math. Helv. 48 (1973), 151–176.

[Rot3] A. Roth: Uniform and tangential approximations by meromorphic functions on closed sets. Can. J. Math. 28 (1976), 104–111.

[Rud] W. Rudin: *Real and complex analysis*. Third Edition. McGraw-Hill, New York, 1987.

[Run] C. Runge: Zur Theorie der eindeutigen analytischen Funktionen. Acta Math. 6 (1885), 228–244.

[Sch] W. J. Schneider: On the growth of entire functions along half-rays. In Entire functions and related parts of analysis, pp.377–385, Amer. Math. Soc., Providence, R.I., 1968.

[Sha1] A. A. Shaginyan: Uniform and tangential harmonic approximation of continuous functions on arbitrary sets. Math. Notes 9 (1971), 78–84.

[Sha2] A. A. Shaginyan: On tangential harmonic approximation and some related problems, Lecture Notes in Mathematics 1275, pp.280–286, Springer, Berlin, 1987.

[Sir] M. Širinbekov: On Hartogs compacts of holomorphy. Math. USSR Sbornik 43 (1982), 403–411.

[Ste] S. Sternberg: *Lectures on differential geometry.* Second Edition. Chelsea, New York, 1983.

[Sze] G. Szegö: *Orthogonal polynomials.* Fourth edition, Amer. Math. Soc., Providence, R.I., 1975.

[VS] G. F. Vincent-Smith: Uniform approximation of harmonic functions. Ann. Inst. Fourier (Grenoble) 19, 2 (1969), 339–353.

[Wal] J. L. Walsh: The approximation of harmonic functions by harmonic polynomials and by harmonic rational functions. Bull. Amer. Math. Soc. (2) 35 (1929), 499–544.

[Zal] L. Zalcman: Uniqueness and non-uniqueness for the Radon transform. Bull. London Math. Soc. 14 (1982), 241–245.

Index

QMW LIBRARY
(MILE END)